T0271018

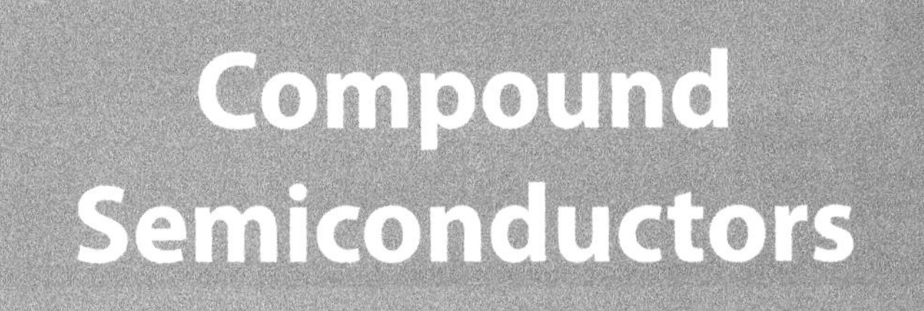
Compound
Semiconductors

Compound Semiconductors

Physics, Technology, and Device Concepts

Ferdinand Scholz

Published by

Pan Stanford Publishing Pte. Ltd.
Penthouse Level, Suntec Tower 3
8 Temasek Boulevard
Singapore 038988

Email: editorial@panstanford.com
Web: www.panstanford.com

British Library Cataloguing-in-Publication Data
A catalogue record for this book is available from the British Library.

Compound Semiconductors: Physics, Technology, and Device Concepts

ISBN 978-981-4774-07-9 (Hardcover)
ISBN 978-1-315-22931-7 (eBook)

Printed in the USA

Contents

Preface

This book is based on a lecture about "Compound Semiconductors," which I have been teaching for many years at Ulm University. It touches a lot of topics which, by themselves, could be and are taken as subjects for other excellent in-depth textbooks. I have indicated such textbooks wherever appropriate to help the interested reader gather more detailed information. However, the main idea of this book is to provide a good general view of this semiconductor family without getting lost in too many details. The reader should acquire all relevant information about this class of semiconductor materials and the technology and processing to synthesize them and fabricate respective device structures. Moreover, the theoretical basis and the most important characterization methods for these materials and structures will be discussed. In contrast to silicon, most compound semiconductors have a direct band structure. That is why they are without any doubt the backbone of the optoelectronics industry. They are also perfectly suited for nanostructure applications. Moreover, their material properties such as carrier mobility, saturation velocity, and breakdown fields allow their use as a basis for high-frequency high-power electronic devices. According to their technical relevance, the major focus is set to the so-called III-V compound semiconductors composed of elements of the third and fifth groups of the periodic system of elements. The major representative of them is GaAs, which is often taken as the leading example of this module. In recent years, group III nitrides with GaN as their most important representative have acquired ever-increasing relevance in all kinds of visible and UV light emission, etc. Hence, a special chapter is devoted to these materials.

The book begins with the presentation of the main principles of semiconductor physics to set a basis for understanding the subsequent chapters. This includes a short recapitulation of crystal lattice basics. Then, the basics of semiconductor band structures are discussed, including the introduction of valence and conduction bands and their most important parameters such as energy gap, density of states, and effective mass. Additionally, carrier statistics and Fermi distribution are addressed.

This general part closes with a short introduction to compound semiconductors.

The chapter on material preparation technology describes the relevant bulk crystal growth methods such as the Czochralski and Bridgman method. The core of this part is devoted to the two modern epitaxial methods—metalorganic vapor phase epitaxy (MOVPE) and molecular beam epitaxy (MBE)—whereas their historical ancestors liquid phase epitaxy (LPE) and hydride vapor phase epitaxy (HVPE) are discussed briefly.

In the Characterization Methods part, we first discuss briefly the electrical properties of semiconductors such as doping, carrier concentration, and mobility. Then, electrical characterization methods such as the van der Pauw Hall-effect method and C-V profiling are considered. The secondary ion mass spectrometry (SIMS) method for the determination of the dopant concentration is also described.

Then, special attention has been paid to optical processes in semiconductors such as absorption and spontaneous and stimulated emission forming the basis for understanding the optoelectronic devices discussed later. Moreover, a wide range of optical spectroscopy methods are presented, including photoluminescence, absorption spectroscopy, and optical gain spectroscopy.

Another important tool to characterize compound semiconductors and respective epitaxial structures is high-resolution X-ray diffraction. Parameters to be measured are lattice constant, strain, composition, superlattice period, and layer quality.

The preceding chapters form a sound base to understand low dimensional structures, which eventually form the core of the devices made of compound semiconductors: quantum wells, wires, and dots. The particular band structure properties of these low-dimensional structures, such as their dimensionality-dependent density of states and energetic eigenstates, are presented in detail. We also discuss the influence of strain on the electronic band structure and its importance in modern laser diodes. Additionally, methods to fabricate quantum wires and quantum dots are presented.

As already mentioned, group III nitrides nowadays require particular attention. Hence a special chapter is devoted to a description of their basic properties—being substantially different to those of the "conventional" arsenides and phosphides—and their fabrication. The specific properties of these materials trigger our discussion of spontaneous and piezoelectric characteristics.

In the last part of the book, we discuss the major device applications of III-V compound semiconductors. According to the technical importance of this field, the chapter on optoelectronic devices has found particular focus in this book. The physics and functionality of optoelectronic device structures such as light-emitting diodes, laser diodes, and solar cells are discussed, which are heavily based on the specific properties of compound semiconductors presented in the preceding chapters of this book. This also includes a description of the most relevant device design and processing details. The book ends with a description of two transistor concepts (field-effect transistor and heterojunction bipolar transistor), again with the main focus on the application and advantages of compound semiconductors in such devices.

I would like to sincerely acknowledge the assistance of all who have helped to realize this book. Many thanks go to Gerlinde Meixner, who has drawn a lot of the figures. Many colleagues and students have contributed by their ideas and inputs. Particularly, many experimental results obtained in my research groups in Stuttgart and Ulm by diploma, master, and PhD students helped to provide instructional diagrams about actual experimental results. Last but not least, I am very grateful to my wife Elke, who assisted this time-consuming project with great patience.

Chapter 1

Basics

This book starts with a brief introduction to semiconductors, including a very short course about solid-state physics, which is helpful to understand the further details. Of course, these fields of science are the subject of many excellent text books. Our description follows to some extent Ref. [1], which contains an excellent in-depth discussion of these topics. See also Chapter 2 of [2].

1.1 What Is a Semiconductor?

As the name indicates, the material class of **semiconductors** differs from other materials by its electrical resistivity ρ, which lies between metals (Cu: $\rho = 1.7 \times 10^{-8}\,\Omega\text{cm}$) and insulators (quartz: $\rho > 10^{21}\,\Omega\text{cm}$). Typical semiconductors span a wide range of resistivity (at least $\rho \sim 10^{-5} \ldots 100\,\Omega\text{cm}$).

Moreover, the following features are typical for semiconductors:

- Their resistivity (or conductivity) can be significantly manipulated by doping.
- Two types of conductivity can be found: n-type and p-type.
- Their conductivity increases with temperature (whereas for metals, it typically decreases!).

Compound Semiconductors: Physics, Technology, and Device Concepts
Ferdinand Scholz

ISBN 978-981-4774-07-9 (Hardcover), 978-1-315-22931-7 (eBook)
www.panstanford.com

Owing to these properties, semiconductors nowadays form the material basis for active electronic devices such as diodes, transistors, and integrated circuits. The technically most important material for these electronic applications is silicon.

However, in this book, we mainly discuss compound semiconductor materials formed by elements of the third and fifth column of the periodic system of elements (PSE). Many of these semiconductors show strong interaction with **light**. Therefore, optoelectronic devices are made of these materials, e.g., light-emitting diodes, laser diodes, photo-detectors, and solar cells.

In the next sections of this chapter, we will briefly review the most important basic properties of semiconductor materials which will enable to understand all these properties mentioned above and will form a sound basis for understanding the functionality of these devices. The first and maybe most important key issue is the so-called **band structure**.

1.2 Naive Band Diagram

Our semiconductors are solid materials forming crystalline structures (see Section 1.3). In a solid, the overlapping electronic orbits of the constituting atoms form bands (see Fig. 1.1). Within these bands, the electrons are no longer bound to one atom, but they are delocalized and (fairly) freely moveable.

The formation of such bands can be understood as follows:

> When two resonant systems with the same resonance frequency ω_0 are (partly) coupled—consider two *LC*-circuits which may be coupled inductively—then the resonance frequency of the resulting system splits into two new frequencies $\omega_0 + \Delta\omega$ and $\omega_0 - \Delta\omega$. The size of $\Delta\omega$ increases with coupling strength. This is also the case if two atoms form a molecule, where chemists are familiar with the two new energy states "HOMO" (highest occupied molecular orbit) and "LUMO" (lowest unoccupied molecular orbit).
> Similarly, the coupling of four resonance systems leads to a coupled system with four new states, where the energy

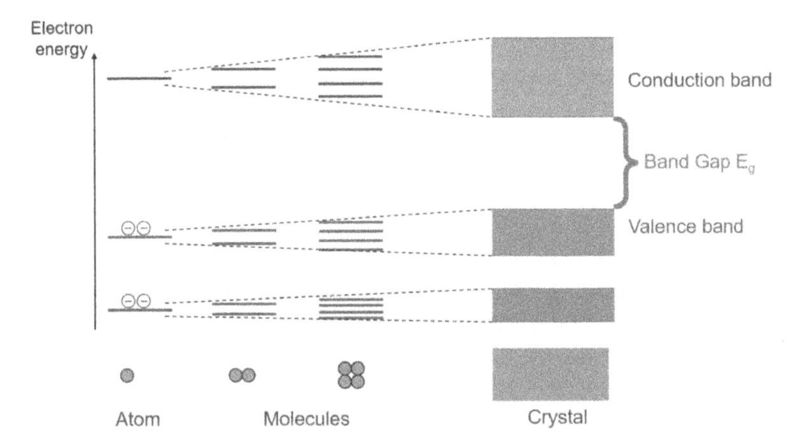

Figure 1.1 From atom to solid. On the left hand side, the atomic orbits are schematically plotted versus energy which then split into new states depending on the number of coupled systems.

separation is smaller than for the system with only two resonators. The energy difference between the new states scales with the reciprocal value of the number of coupled states. Consequently, the coupling of very many atoms in a crystal leads to a new system with very many energy states built from the equivalent orbits of each atom, which are only separated by a very small energy gap, much smaller than what can be distinguished (particularly on a scale of thermal energy kT, even for very small temperatures). Hence roughly each atomic orbital leads to the formation of a band in which the carriers can move fairly freely and are no longer localized to a single atom (for more details, see below).

The most important bands are

- valence band (highest band occupied with electrons), mainly formed by the valence electrons of the atoms forming the crystal;
- conduction band (lowest empty band), formed by the atomic states energetically directly above the states of the valence electrons.

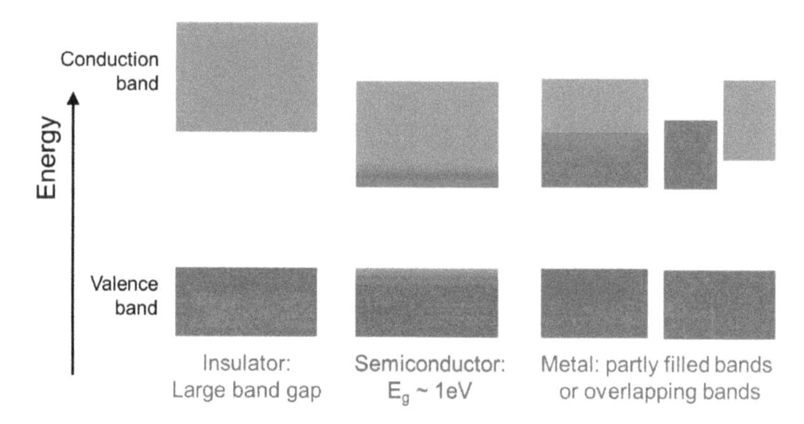

Figure 1.2 Insulators, metals, semiconductors schematically: Relation to band structures.

In semiconductors, these two bands are separated by an energy difference of some 100 meV to a few eV. This difference is called **band gap** E_g. As it is directly responsible for many basic properties of a semiconductor, it may be regarded as the key property of a semiconductor.

Note: Bands fully occupied by carriers do not contribute to electrical current. However, partly occupied bands may carry some current. This will be explained in more detail later in this chapter.

This already enables us to identify the type of conductance of a crystalline material by its band structure (see Fig. 1.2):

- Insulators: The valence band is completely filled with carriers, conduction band is completely empty. Therefore, any movement of carriers within one band does not change the total situation (see Section 1.6). Consequently, no net electrical current conduction can occur, and the material is insulating.
- Metals: **Either** the valence band is only partly filled (e.g., elements of groups I and III of PSE like Na or Al) **or** the two highest bands overlap (like for elements of group II like Zn, Mg). The number of conducting electrons is roughly the same as the number of atoms. Now, an electric field will

result in a rearrangement of the carriers in a band, leading to a net electrical current conduction.

- Semiconductors: At low temperature they are in the same situation as insulators: The valence band is completely occupied with carriers, whereas the conduction band is completely empty. However, the distance between valence and conduction band (called **band gap** E_g) is fairly small as measured on the scale of thermal energy kT where k is Boltzmann's constant ($k = 1.38 \times 10^{-23}$ J/K) and T is the absolute temperature. Therefore, at finite temperature, particularly at room temperature, electrons can be thermally excited from the valence into the conduction band. Hence, both bands are no longer completely full or completely empty (Fig. 1.2), thus causing some electrical conductivity. Typical band gaps are 0.67 eV for Ge, 1.14 eV for Si, or 1.42 eV for GaAs. However, the number of thermally excited carriers (or of carriers generated by doping, see later) is typically much smaller than that of the atoms. Therefore, semiconductors have a much lower conductivity as compared to metals.

1.3 Crystal Lattice: Basics

Most solids (and thus most semiconductors) form crystals, i.e., very regular arrangements of their constituting atoms. The crystal structure is responsible for many basic material properties.

Whether a solid-state material is a crystal, is defined by its symmetry. Each crystal possesses at least translation symmetry:

$$\vec{R}' = \vec{R} + n_1 \cdot \vec{a} + n_2 \cdot \vec{b} + n_3 \cdot \vec{c}, \tag{1.1}$$

where $\vec{R}'$ and $\vec{R}$ are position vectors and n_i are integers. The vectors $\vec{a}$, $\vec{b}$, $\vec{c}$ are linear independent vectors.

Each lattice point defined by this equation may have a specific structure built by the arrangement of several atoms, forming the basis of the lattice.

Hence, Lattice + Basis = Crystal structure (see Fig. 1.3).

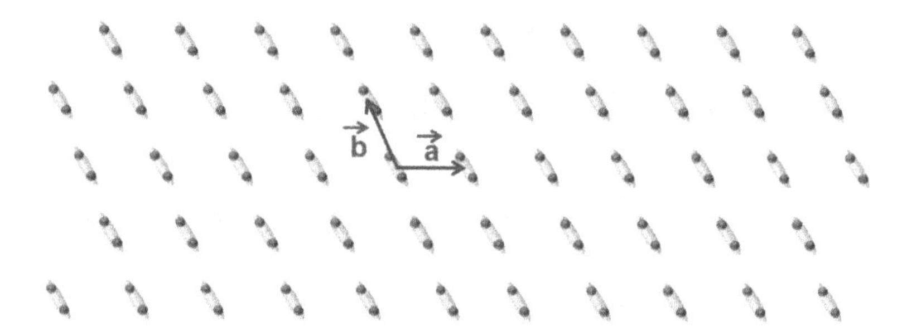

Figure 1.3 Lattice composed by complex basis units (schematically).

The number of atoms forming the basis depends on the specific crystal. It may be

- just one atom: Metals, frozen noble gases (as solids);
- several atoms: many crystals including compound semiconductors;
- thousands of atoms: organic molecule crystals.

A crystal can be regarded as being formed by unit cells, which fill the space totally, if translation is applied appropriately. The translation vectors to move from one to the next unit cell (see Eq. 1.1) and in particular their magnitudes are called **lattice constants**.

A primitive unit cell contains just one basis (one lattice point) of the crystal. However, other unit cells can be defined which may not necessarily be primitive, but more appropriate to describe the crystal! See some examples in Fig. 1.4.

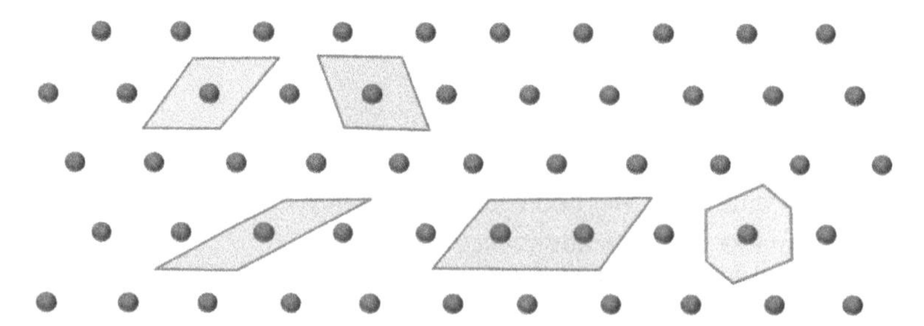

Figure 1.4 Differently defined unit cells containing one (left) or two (center) lattice points. Right: Wigner–Seitz cell, see also Fig. 1.5.

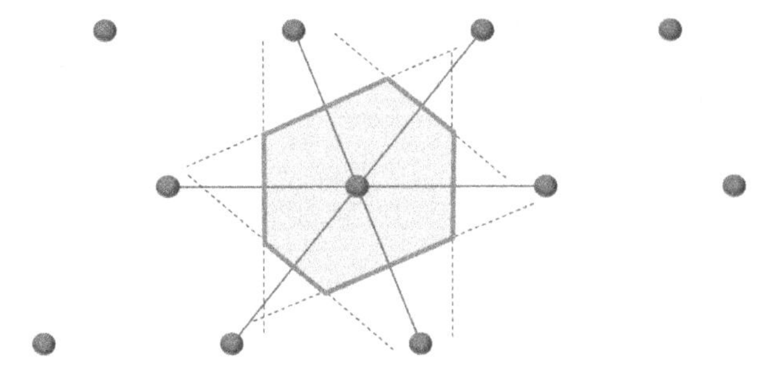

Figure 1.5 Construction of the Wigner–Seitz unit cell for a two-dimensional lattice.

One type of primitive unit cell is the **Wigner–Seitz cell**, which can be found by the following procedure:

- Form midplanes on connecting straight lines from one lattice point to all others (only nearer neighbors are of larger importance, see Fig. 1.5).
- Then, the smallest volume around the central lattice point formed by these midplanes is called the **Wigner–Seitz cell**.

Besides translation operation, also other symmetry operations are typical for crystals and may be used to classify them:

- Rotation by $2\pi/2$, $2\pi/3$, $2\pi/4$, $2\pi/6$
- Reflection
- Inversion (reflection + rotation)

From these symmetry properties, 14 different point lattices can be deduced, the so-called **Bravais lattices** (see Table 1.1) which are different with respect to their allowed symmetry operations, etc.

Some crystal structures—being important for compound semiconductors—are a consequence of the densest package of bowls (Fig. 1.6).

When you stack planes of bowls (think, e.g., about oranges presented at a fruit market), then the second plane falls into the voids of the first plane. However, only every second void can be occupied. Therefore, the bowls in the third plane have two basically different options:

Table 1.1 Basic characteristics of the 14 Bravais lattices.

Symmetry	Lattice constants	Angles	Sub-types
Cubic	$a = b = c$	$\alpha = \beta = \gamma = 90^\circ$	P, BC, FC
Tetragonal	$a = b \neq c$	$\alpha = \beta = \gamma = 90^\circ$	P, BC
Orthorhombic	$a \neq b \neq c$	$\alpha = \beta = \gamma = 90^\circ$	P, BC, FC, SC
Hexagonal	$a = b \neq c$	$\alpha = \beta = 90^\circ, \gamma = 120^\circ$	P
Trigonal	$a = b = c$	$\alpha = \beta = \gamma \neq 90^\circ$	P
Monoclinic	$a \neq b \neq c$	$\alpha = \beta = 90^\circ, \gamma \neq 120^\circ$	P, SC
Triclinic	$a \neq b \neq c$	$\alpha, \beta, \gamma \neq 90^\circ$	P

Note: P = primitive, BC = body centered, FC = face centered, SC = side centered.

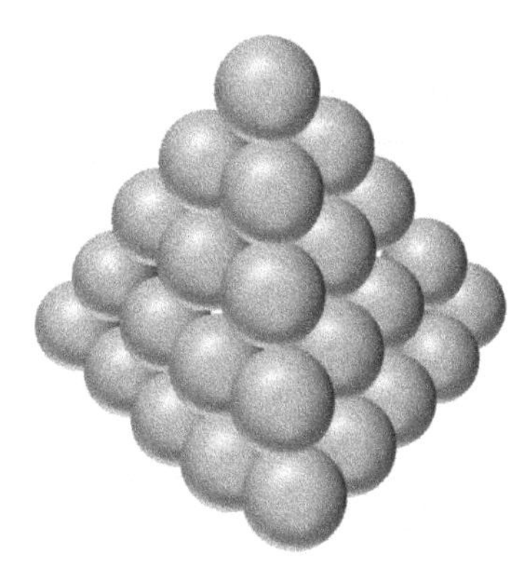

Figure 1.6 Densest package of bowls.

- They fall into the voids of the second plane which are above the not used voids of the first plane (Fig. 1.7, left). This leads to a stacking sequence A-B-C-A-B-C and consequently to cubic crystal symmetry. Our elemental semiconductors (Ge, Si, C) and most compound semiconductors discussed in this book such as GaAs, InP, and many others form such crystals. This structure may be regarded as most important for technically relevant semiconductors. It results in the face-centered cubic (fcc) crystal structure (see Fig. 1.8).
  Note: For such materials, we use a cube as unit cell which, however, is **not** a primitive unit cell, see Fig. 1.9. It contains four lattice points!

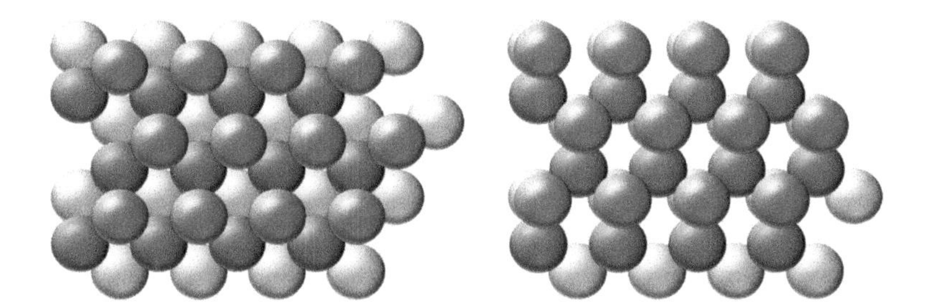

Figure 1.7 Stacking of cubes seen from top. Left: Sequence A-B-C; right: Sequence A-B-A.

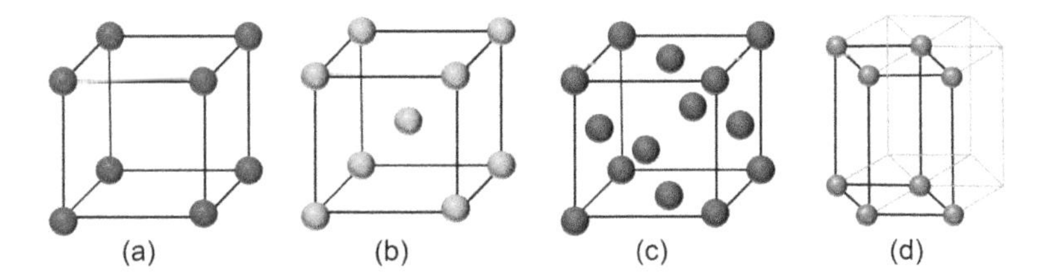

Figure 1.8 The primitive (a), body centered (b), face centered cubic (c), and the hexagonal lattice (d).

Moreover, The basis is constructed by two atoms, even for the **diamond** structure, in which C, Si, and Ge crystallize: Both atoms are the same. If the two atoms are different, then this structure is called **zinc blende**. This is the case for our cubic compound semiconductors like GaAs and InP.
One atom of the basis may occupy the lattice points (e.g. corners of the cube and the 6 face centers), the second atom is then shifted by 1/4 lattice constant a in all three directions (occupying the position which may be described by the vector $(a/4, a/4, a/4)$).

- The other option is that the bowls of the third plane fall into the voids which are above the bowls of the first plane (Fig. 1.7, right). This leads to the stacking sequence A-B-A-B-A-B and hence to a hexagonal crystal symmetry. Particularly GaN and its related compounds (see Chapter 11), SiC, and many II-VI compound semiconductors (formed by elements of the second and sixth column of the PSE) form such

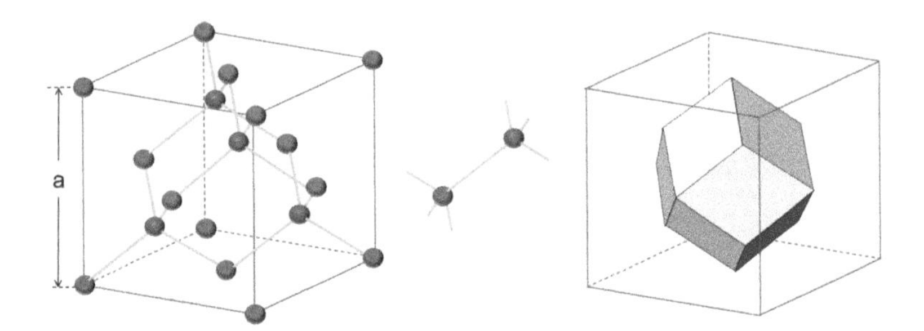

Figure 1.9 The zinc blende lattice (fcc), which is identical to the diamond lattice (for C, Si; and Ge), if both atoms of the primitive unit are the same type. Left: Conventional cubic unit cell; center: 2-atomic unit forming the basis of the periodic structure; right: Wigner–Seitz cell containing one such unit.

crystals. A prominent example is ZnO, giving the name to this class of crystal symmetry: **Wurtzite** crystal structure which belongs to the hexagonal lattice class (see Fig. 1.8).

We will later discuss the consequences of this seemingly minor difference.

1.3.1 Crystal Planes, Miller Indices

In order to specify different crystal orientations, crystal planes can be identified by three Miller indices. They can be found by the following procedure (Fig. 1.10):

(1) Find intercepts of plane with the three basic crystal axes in terms of the lattice constants.
(2) Take reciprocals of these numbers.
(3) Reduce them to the smallest three integers with same ratio.

Some conventions:

- Miller indices for a single plane or set of parallel planes: Indices enclosed in brackets: (hkl).
- A negative index (when the plane intercepts on the negative side) is typically indicated by a bar above the number: $(0\bar{1}1)$

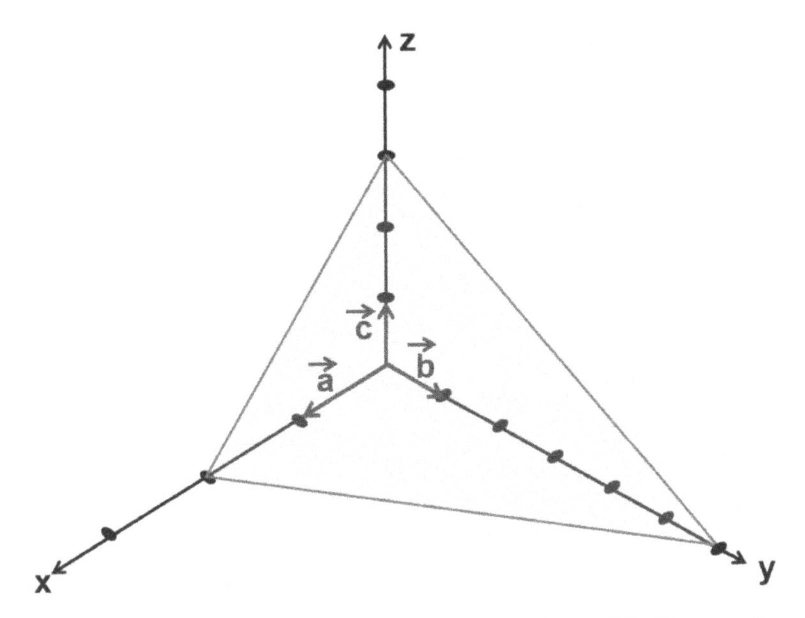

Figure 1.10 How to find Miller's indices for a crystal plane? In this example, the plane intercepts the x-axis at 2 lattice units, the y-axis at 6 units, and the z-axis at 3 units. Hence its Miller's indices are (6 2 4).

- Planes of equivalent symmetry like (100), (010), (001), $(\bar{1}00)$, $(0\bar{1}0)$, $(00\bar{1})$ in cubic crystals (the 6 planes forming the cube) are enclosed in curly brackets: {100}.
- A direction in a crystal is defined by a vector in units of the three base vectors. It is also indicated by three indices, but in square brackets: [010]. A set of symmetrically equivalent directions is given by angle brackets: <010>.

In cubic crystals, directions are perpendicular to the planes with the same Miller indices. However, this may not be the case in other Bravais lattices.

1.3.2 Reciprocal Lattice

Lattice properties may be measured by x-ray diffraction, because the wavelength λ of x-rays is in the same range as the lattice constants. Typically, x-ray diffraction is used for this purpose.

If a monochromatic (single wavelength) plane x-ray wave falls onto a crystal, it gets reflected at the crystal planes (see Fig. 1.11). Only if the beams reflected at each crystal plane interfere constructively, an outgoing reflected beam can be observed. This is described by Bragg's law for the constructive reflection of an x-ray beam with an angle of incidence of ϑ on parallel lattice planes having a spacing d (Fig. 1.11):

$$n \cdot \lambda = 2d \sin \vartheta, \tag{1.2}$$

where the integer number n describes the order of constructive interference.

This can be also derived from a seemingly very different description: Laue considered how an incoming x-ray wave $\psi = \psi_0 \cos(\omega t - \vec{k}\vec{r})$ with wave vector $\vec{k}$ interacts with periodically arranged lattice points (Fig. 1.12): We only consider elastic interaction, i.e. $\omega' = \omega$ (conservation of energy) and therefore $|\vec{k}'| = |\vec{k}|$.

At each lattice point, spherical waves are generated. The diffracted wave ψ' will only go out in this direction, where all these spherical waves interfere constructively. It may have another direction: $\vec{k}' \neq \vec{k}$. It can be easily shown that now the condition for constructive interference reads

$$\Delta\vec{k} = \vec{k}' - \vec{k} \qquad \text{conservation of momentum} \tag{1.3}$$

with

$$\vec{a} \cdot \Delta\vec{k} = 2\pi h \tag{1.4}$$

$$\vec{b} \cdot \Delta\vec{k} = 2\pi k \tag{1.5}$$

$$\vec{c} \cdot \Delta\vec{k} = 2\pi l. \tag{1.6}$$

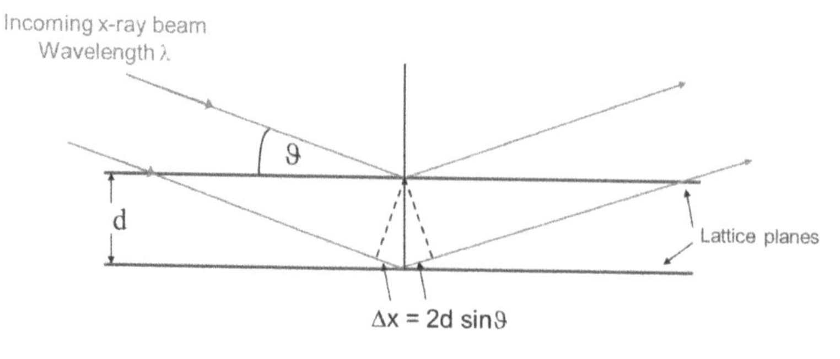

Figure 1.11 Bragg's law of diffraction.

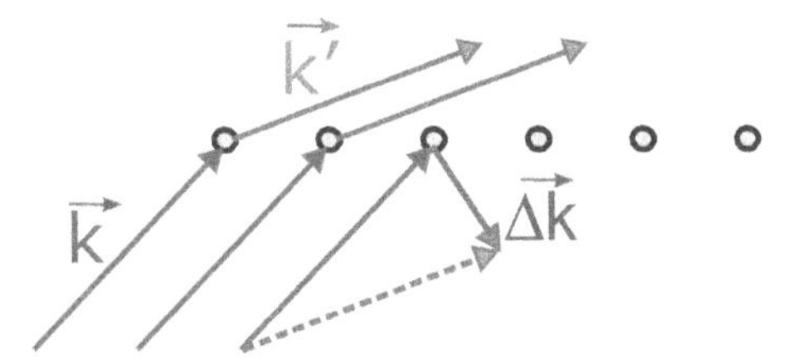

Figure 1.12 Interaction of a x-ray wave with a point lattice according to Laue's description

These are Laue's equations with $\vec{a}$, $\vec{b}$, $\vec{c}$: Lattice vectors and h, k, l: integers. They are completely equivalent to Bragg's equation 1.2.

This can be also written as

$$\Delta\vec{k} = h \cdot \vec{A} + k \cdot \vec{B} + l \cdot \vec{C}, \tag{1.7}$$

with the reciprocal lattice vectors

$$\vec{A} = 2\pi \frac{\vec{b} \times \vec{c}}{\vec{a} \cdot \vec{b} \times \vec{c}}$$

$$\vec{B} = 2\pi \frac{\vec{c} \times \vec{a}}{\vec{a} \cdot \vec{b} \times \vec{c}}$$

$$\vec{C} = 2\pi \frac{\vec{a} \times \vec{b}}{\vec{a} \cdot \vec{b} \times \vec{c}}$$

and h, k, l being integers.

Hence, any arbitrary reciprocal lattice vector $\vec{G}$ is formed by combining these base vectors:

$$\vec{G} = h \cdot \vec{A} + k \cdot \vec{B} + l \cdot \vec{C} \tag{1.8}$$

Please notice that Eq. 1.8 is completely equivalent to Eq. 1.1, hence defining some crystal lattice. Indeed, by these vectors, the fundamental translations of the so-called **reciprocal lattice** are defined. This reciprocal lattice is the three-dimensional Fourier transformation of the original lattice. Therefore, it is another complete description of our original lattice. It has a big importance in solid-state physics and will be heavily used in the subsequent parts of this book.

This reciprocal lattice has all properties of a real lattice. In particular, we can also define some appropriate unit cells. The most primitive unit cell, the Wigner–Seitz cell of the reciprocal lattice is called **Brillouin zone**. Please keep in mind: The primitive unit cell contains (more or less) all properties of a crystal.

The symmetries of the reciprocal lattice are a consequence of the symmetries of the real lattice. In particular, one can show: If the original lattice is orthogonal, then also the reciprocal lattice is orthogonal. Moreover, the following relations hold between the different cubic lattices:

Original lattice	Reciprocal lattice
C	C
FCC	BCC
BCC	FCC

1.4 More Details about Band Structure

In this section, we will briefly discuss some basics about solid-state quantum physics. For a full description, we refer the reader to appropriate textbooks.

The main goal is to get a basic understanding of the electronic states, movements, etc., of carriers in a solid which is closely related to their quantum mechanical wave function Ψ with the two parameters **energy** E and **momentum** $\vec{k}$.

Remember the classical relation between the kinetic energy E of a particle and its momentum $\vec{p}$:

$$E = \frac{1}{2}mv^2 = \frac{p^2}{2m} \qquad \text{with} \qquad p = mv \tag{1.9}$$

The momentum of a quantum mechanical particle is given by: $\vec{p} = \hbar\vec{k}$.

Therefore, we find for a freely moving electron:

$$E = \frac{\hbar^2 k^2}{2m} = \frac{\hbar^2}{2m}(k_x^2 + k_y^2 + k_z^2) \tag{1.10}$$

However, in a solid, an electron is not completely free but interacts with the atoms of the crystal. These atoms form a periodic potential $V(\vec{r}) = V(\vec{r} + \vec{R})$ where $\vec{R}$ is a lattice vector (according to Eq. 1.1).

In many cases, this potential is only a small perturbation, which leads us to the concept of a nearly free electron (or more generally, a free carrier).

The quantum-mechanical particle wave function has to fulfill the Schrödinger equation.

1.4.1 Insert: Very Short Introduction to Quantum Mechanics

Basic idea: Many things are quantized, i.e., they cannot be divided into infinitely small parts.

Example: The quantum of photon energy is $h\nu$.

This requires a completely new description of physics!

Basic assumptions of quantum physics:

(1) The state of motion of a particle can be described by its wave function $\Psi(\vec{r}, t)$. This wave function must be so that a normalization is possible:

$$\int_{-\infty}^{\infty} |\Psi(\vec{r}, t)|^2 d\vec{r} = 1$$

(2) Every measurable quantity (e.g., position or momentum of the particle) corresponds to a Hermitian operator A. The expected value of a measurement for the observable corresponding to A is

$$< A >= \int_{-\infty}^{\infty} \Psi^* A \Psi d\vec{r}$$

(3) Some examples for such operators:

- Position: $A = x$
- Momentum: $A = p = -i\hbar\frac{d}{dx}$ (one dimension) or $-i\hbar(\frac{\partial}{\partial x}, \frac{\partial}{\partial y}, \frac{\partial}{\partial z}) = -i\hbar\nabla$ (three dimensions).

(4) The result of a measurement yields an eigenvalue λ of the operator A, i.e., the equation (in most cases a differential equation)

$$A\Psi_k = \lambda\Psi_k$$

must be fulfilled for some scalar λ and some wave function Ψ_k.

(5) The fundamental law of motion of quantum physics is the Schrödinger equation

$$H\Psi = i\hbar\frac{\partial\Psi}{\partial t} \qquad \text{time dependent} \tag{1.11}$$

$$H\Psi = E\Psi \qquad \text{time independent,} \tag{1.12}$$

where H is the operator of the classical Hamiltonian function which is related to the law of energy conservation:

$$H \equiv E_{\text{kin}} + E_{\text{pot}} = E_{\text{total}}$$

With the operator for the momentum as given above, we get the operator for the kinetic energy as

$$\frac{p^2}{2m} = \frac{1}{2m}(i\hbar)^2\left(\frac{\partial}{\partial x}, \frac{\partial}{\partial y}, \frac{\partial}{\partial z}\right)\cdot\left(\frac{\partial}{\partial x}, \frac{\partial}{\partial y}, \frac{\partial}{\partial z}\right)$$
$$= -\frac{\hbar^2}{2m}\left[\frac{\partial^2}{\partial x^2} + \frac{\partial^2}{\partial y^2} + \frac{\partial^2}{\partial z^2}\right] = \frac{-\hbar^2}{2m}\triangle$$

We can deduce the Schrödinger equation also by some other way:

Quantum physics tells us that not only light waves need to be described by particle properties sometimes, but also a particle with a mass m and a velocity $\vec{v}$ may be described by a wave with frequency ν: **matter wave**.

A plane wave is described by

$$A(\vec{r}, t) = A_0 e^{j(\omega t - \vec{k}\vec{r})}, \tag{1.13}$$

which fulfills the general wave equation

$$\triangle A(\vec{r}, t) = \alpha\frac{\partial^2 A(\vec{r}, t)}{\partial t^2}.$$

For a particle, we may define the following:

Energy $E = h\nu = \hbar\omega$ with $h = 2\pi\hbar = 6.626 \times 10^{-34}$ Js (Planck's constant or quantum of action).

Similarly, the vector of momentum reads $\vec{p} = m\vec{v} = \hbar\vec{k}$. Thus, the wave function reads

$$\Psi(\vec{r}, t) = \Psi_0 e^{\frac{j}{\hbar}(Et - \vec{p}\vec{r})}. \tag{1.14}$$

The total energy is $E = E_{\text{kin}} + E_{\text{pot}}$ with $E_{\text{kin}} = \frac{1}{2}mv^2 = \frac{p^2}{2m}$, which can be multiplied with Ψ:

$$\left(\frac{p^2}{2m} + E_{\text{pot}}\right)\Psi = E\Psi \tag{1.15}$$

From Eq. 1.14, we get

$$\frac{d\Psi}{dt} = j\frac{E}{\hbar}\Psi \qquad \Rightarrow \qquad E\Psi = -j\hbar\frac{d\Psi}{dt} \tag{1.16}$$

and

$$\Delta\Psi = -\frac{p^2}{\hbar^2}\Psi \qquad \Rightarrow \qquad \frac{p^2}{2m}\Psi = -\frac{\hbar^2}{2m}\Delta\Psi. \tag{1.17}$$

Both inserted in Eq. 1.15 yields

$$\left(-\frac{\hbar^2}{2m}\Delta + E_{\text{pot}}\right)\Psi = -j\hbar\frac{d\Psi}{dt} = E\Psi, \tag{1.18}$$

which is again the Schrödinger equation.

1.4.2 Schrödinger Equation and Dispersion Relation

Hence, for our quasi-free electron in a crystal, the following Schrödinger equation is valid:

$$\left[-\frac{\hbar^2}{2m}\left(\frac{\partial^2}{\partial x^2} + \frac{\partial^2}{\partial y^2} + \frac{\partial^2}{\partial z^2}\right) + V(\vec{r})\right]\Psi_k(\vec{r}) = E_k\Psi_k(\vec{r}) \tag{1.19}$$

The solutions which we search should be valid in a large, if not even infinite crystal. Such a situation can be best considered by assuming periodic boundary conditions, i.e., the crystal is repeated periodically with a (macroscopic) period L. In three dimensions, we may consider a cube with length L.

For our wave function, this requires

$$\Psi(nL, t) = \Psi(0, t),$$

where n is an integer number.

Then, a wave function which describes best this problem is a standing wave with

$$\Psi_n(\vec{r}) = A_n \sin\frac{2n_x\pi x}{L}\sin\frac{2n_y\pi y}{L}\sin\frac{2n_z\pi z}{L},$$

with $n_i = 1, 2, 3, 4, \ldots, \infty$.

Thus the momentum for these periodic boundary conditions is $k_i = \frac{2n_i\pi}{L}$.

If the potential is periodic, then not all n are allowed (as a consequence of the Bloch theorem[a]).

Because of the translation symmetry of the reciprocal lattice, we know: If a solution $\vec{k_0}$ is found, then $\vec{k_0} + \vec{G}$ is another solution.

For simplification, we consider just a linear chain of atoms, i.e. a one-dimensional crystal with lattice constant a.

Then a periodic wave function is $\Psi_n(x) = A \sin(2\pi nx/L) = A \sin kx$.

If $k = \frac{2\pi}{a}$, i.e., $\lambda = a$, then any wave reflected at one of the atoms interferes constructively. This is equivalent to Bragg's law:

$$(\vec{k} + \vec{G})^2 = k^2$$

or in one dimension,

$$k = \pm\frac{1}{2}G = \pm\frac{n\pi}{a}.$$

For this k-vector, we get reflection of the electron wave and thus a standing wave which may be

$$\begin{aligned} &\text{either} \quad \Psi(+) \sim \cos\left(\frac{\pi x}{a}\right) \\ &\text{or} \quad \Psi(-) \sim \sin\left(\frac{\pi x}{a}\right). \end{aligned}$$

It is evident that these functions have different mean distance from the atoms, therefore they possess different potential energy. This gives rise for an energy separation at $k = \pm n\frac{\pi}{a}$ (see Fig. 1.13). This is another explanation of the band gap E_g!

Moreover, any periodic wave with $k > \frac{\pi}{a}$ is equivalent to a periodic wave with $k' = k - G$ where, G is a reciprocal lattice vector.

This means we can map all waves with $k > \frac{\pi}{a}$ onto the first Brillouin zone where $k \leq \frac{\pi}{a}$ (Fig. 1.13, right).

---

[a] Here the Bloch theorem is important: If the potential energy E_{pot} is periodic (as it is in a crystal) with the periodicity of the lattice, i.e., $E_{\text{pot}}(\vec{r}) = E_{\text{pot}}(\vec{r} + \vec{R})$ with $\vec{R}$ being a lattice vector, then the solutions of the Schrödinger equation are of the form

$$\Psi(\vec{k}, \vec{r}) = e^{j\vec{k}\vec{r}} U_n(\vec{k}, \vec{r}),$$

where $U_n(\vec{k}, \vec{r})$ is periodic in $\vec{r}$ with the periodicity of the lattice

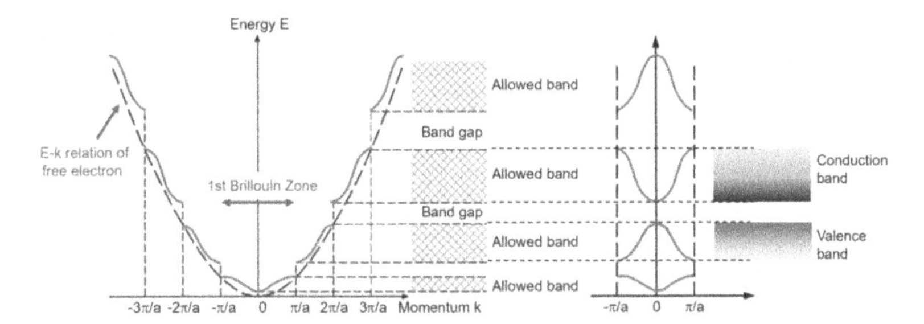

Figure 1.13 Left: E-k relationship for electrons subjected to the potential distribution of the Kronig–Penney model and the corresponding energy band structure. Please notice the change of the parabola of the free electron by the periodic potential thus forming the band gaps. Right: Simplified band structure by back-folding of higher bands into the first Brillouin zone.

1.4.3 Short Summary Up to Now

- Electrons are distributed in bands. Most important: **Valence band** (highest filled band) and **conduction band** (lowest empty band).
- Electron movement described by wave function and then by dispersion relation $E(\vec{k})$.
- $\vec{k}$ can always be projected into first Brillouin zone by adding an adequate reciprocal lattice vector $\vec{G}$.
- We have discussed two demonstrative explanations for the band gap E_g which separates the valence and the conduction band:
  - from atom to solid: formation of bands from coupling of atomic levels
  - standing waves in a crystal due to reflection for $\vec{k} = n \cdot \vec{G}$

1.5 Effective Mass

Within the bands, the electrons can move as quasi-free particles. This concept is even closer to a really free electron, if all interaction with the lattice potential is taken into account by ascribing an **effective mass** m^* to this particle which differs from the free electron mass m_0. How is this mass related to the band structure?

Let us again describe the electron with a wave function $\Psi(\vec{r}, \vec{t})$ with wave vector $\vec{k}$ and frequency ω (see Eq. 1.13). The group velocity of such a wave is

$$v_{\mathrm{gr}} = \frac{d\omega}{dk}. \tag{1.20}$$

With $E = \hbar\omega$ we get

$$v_{\mathrm{gr}} = \frac{1}{\hbar}\frac{dE}{dk}. \tag{1.21}$$

Then the acceleration reads

$$a = \frac{dv_{\mathrm{gr}}}{dt} = \frac{1}{\hbar}\frac{d}{dt}\frac{dE}{dk} = \frac{1}{\hbar}\left(\frac{d^2E}{dk^2}\right)\frac{dk}{dt}. \tag{1.22}$$

If we apply an external electric field F, then the electron takes up some energy dE during the time interval dt:

$$dE = eFv_{\mathrm{gr}}dt = \frac{eF}{\hbar}\frac{dE}{dk}dt. \tag{1.23}$$

With $dE = \frac{dE}{dk}dk$ we get

$$\frac{dk}{dt} = \frac{eF}{\hbar} \tag{1.24}$$

and thus

$$a = \frac{eF}{\hbar^2}\left(\frac{d^2E}{dk^2}\right). \tag{1.25}$$

Remember: Force$= eF = ma$, therefore $a = \frac{eF}{m}$. Thus finally we see

$$m^* = \hbar^2\left(\frac{d^2E}{dk^2}\right)^{-1}. \tag{1.26}$$

i.e., the effective mass is inversely proportional to the curvature of the dispersion relation.

We also notice: At least for small k, i.e., in the center of the first Brillouin zone, the dispersion relation can be approximated by a parabola. Then it reads

$$E(k) = \frac{\hbar^2k^2}{2m^*} \tag{1.27}$$

similar as for free electrons, but with effective mass m^* describing the influence of the lattice on the electron movement. Typical values for the effective mass are $m^* = 0.57m_0$ for Ge, $m^* = 0.063m_0$ for GaAs, or $m^* = 0.2m_0$ for GaN, where m_0 is the rest mass of the electron ($m_0 = 9.11 \times 10^{-31}$ kg).

1.6 Movement of Carriers: Concept of Hole

A quasi-free electron in a band in an electric field F takes up some energy and thus changes its wave vector $\vec{k}$ by $\Delta\vec{k}$. If it reaches the edge of the first Brillouin zone (BZ), then it is scattered back to the other end of the BZ ($\vec{k'} = \vec{k} - \vec{G}$). Therefore, if all places are occupied in a band, then all electrons change by $\Delta\vec{k}$. Those which are back-scattered to the other end of the Brillouin zone just occupy the states which have become vacant there, because the original electrons here are shifted by $\Delta\vec{k}$. This means that finally no macroscopic change occurs in such a band, i.e., no electrical conduction occurs.

If a band is partly filled, then all electrons take up a momentum $\Delta\vec{k}$. There is no compensation by back-scattered electrons. Therefore, the center of gravity of momentum of the electrons changes in an electric field, which is equivalent to some electrical conduction.

An electron may be excited from the valence band into the conduction band leaving a vacancy in the conduction band. If now an electric field is applied, the remaining electron ensemble in the valence band changes its momentum again. This is equivalent to the movement of the vacancy in the opposite direction. In summary this vacancy behaves like a positively charged particle. This is a quasi-particle called **hole**. At the top of the valence band (negative curvature) an electron would have negative mass (see Eq. 1.26). Hence, a positive mass can be ascribed to the positively charged quasi-particle **hole**! We will later see that with the help of this quasi-particle, the description of our semiconductor material and device properties gets significantly simpler.

1.7 Real Band Structures

Figure 1.13 shows some simplified and idealized dispersion relation of carriers in crystals. In reality, such relations, i.e., the real band structures, look somewhat more complicated (see Fig. 2.1). Here, we consider two (still simplified) cases:

(1) In a **direct semiconductor**, the minimum of the conduction band and the maximum of the valence band are in the center

of the first Brillouin zone (BZ) at $k = 0$ (Fig. 1.14, left). This is true for many compound semiconductors such as GaAs, InP, and ZnSe. These semiconductors interact strongly with light (see Section 1.9). Therefore, they are well suited for light-emitting devices such as LEDs and laser diodes.

(2) If the minimum of the conduction band is not at $k = 0$ (the maximum of the valence band is always at $k = 0$), then it is an **indirect semiconductor** like Si or Ge (Fig. 1.14, right). These materials interact only weakly with light. That is the reason why they cannot be used in light-emitting devices (LEDs, lasers).

The conduction band minimum is typically isotropic, i.e., with spherical energy surface, as it stems from the *s* orbitals of the atoms.

The valence band structure is typically more complicated than the conduction band structure (Fig. 1.15). Originally it is triply degenerate at $k = 0$, as it stems mainly from the *p* orbitals of the atoms. By the spin-orbit interaction, one valence band is split off even at $k = 0$ ("split-off band"). The two remaining top-most bands have different curvature, hence resulting in a *heavy hole* and a *light hole* band (see Fig. 1.15), which are degenerate (i.e., they have the same energy) at $k = 0$.

In a three-dimensional crystal, the dispersion relations may be different for the different directions $\vec{k}_x$, $\vec{k}_y$ and $\vec{k}_z$. This makes the

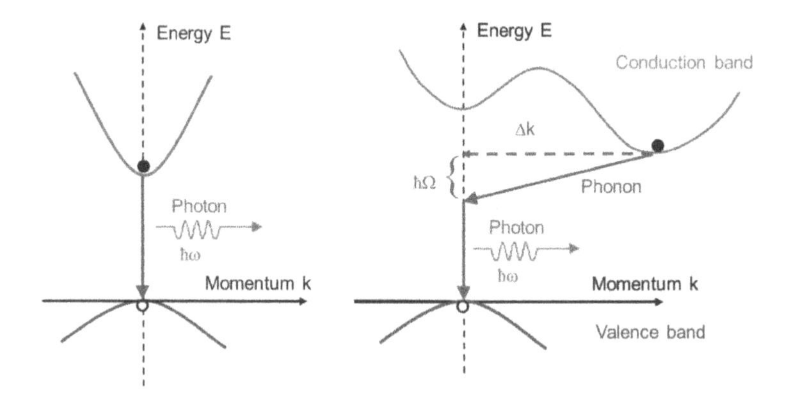

Figure 1.14 Direct (left) and indirect (right) band structure. In the latter, another particle (typically a phonon) is required for the conservation of momentum in the emission or absorption process of a photon.

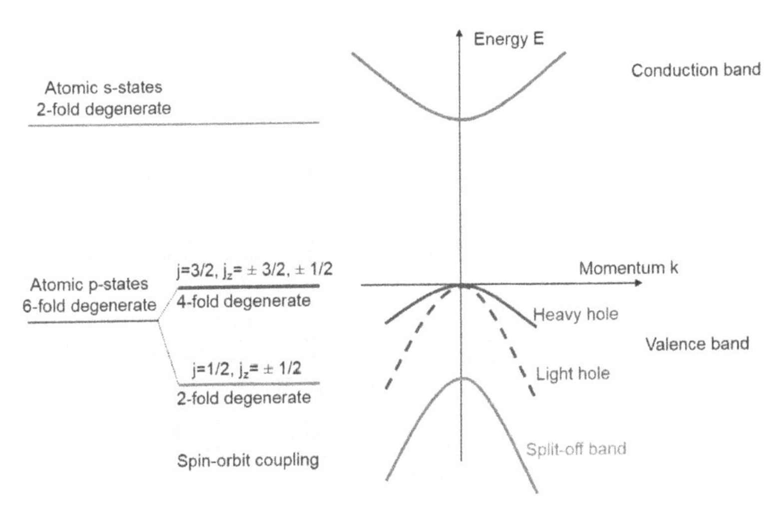

Figure 1.15 Details of a typical valence band of (cubic) compound semiconductors.

band structure more complex (see different axes in Fig. 2.1). Some details will be discussed later.

1.8 Temperature Dependence of Lattice Constant and Band Gap

As expected, most, if not all characteristics of semiconductors change with temperature. This is particularly the case for the lattice constant, which increases with temperature similar as well-known for other materials. This behavior is typically described by

$$a(T) = a_0(1 + \alpha T), \tag{1.28}$$

where $a(T)$ is a characteristic length (e.g., the lattice constant) at a given temperature T and a_0 is this length at a fixed temperature (e.g., room temperature). The thermal expansion coefficient α depends slightly on temperature but is very often assumed to be constant. For compound semiconductors, a typical value is $\alpha \sim 5 \times 10^{-6}\,\mathrm{K}^{-1}$. The thermal expansion may become important in heterostructures causing some temperature-induced strain, if the various layers have different thermal expansion coefficients (see Chapter 9).

The band gap $E_g(T)$, however, shrinks with temperature with respect to the band gap $E_g(0)$ at $T = 0$ K. Typically, this is described with Varshni's formula [3]:

$$E_g(T) = E_g(0) - \frac{AT^2}{B+T} \tag{1.29}$$

with the parameters A and B. The latter can be identified as **Debye temperature**.

The band gap difference between $T = 0$ K and room temperature amounts typically to values around 70–100 meV.

It is interesting to note that both features are related: A smaller lattice constant typically leads to a larger band gap. This can be seen, besides in these thermal effects, when looking to various semiconductors with different lattice constants, see, e.g., Fig. 2.3. The band gap also increases if the lattice constant shrinks by (internal or externally applied) strain (see Chapter 9).

1.9 Interaction with Light

The excitation of carriers from the valence band (VB) to the conduction band (CB) may be initiated by light: When a photon of respective energy $h\nu > E_g$ is absorbed, the energy may be transferred to an electron which thus is excited from the VB to the CB.

The amount of light intensity I_T penetrating some material of thickness d is described by the absorption coefficient α of this material:

$$I_T = I_0 e^{-\alpha d},$$

where I_0 describes the original light intensity. The absorbed light intensity I_a is then given by

$$I_a = I_0 - I_T.$$

The excitation (or the recombination) process can be regarded as a collision process of two particles: The photon collides with the electron thus transferring its energy to the electron (or vice versa).

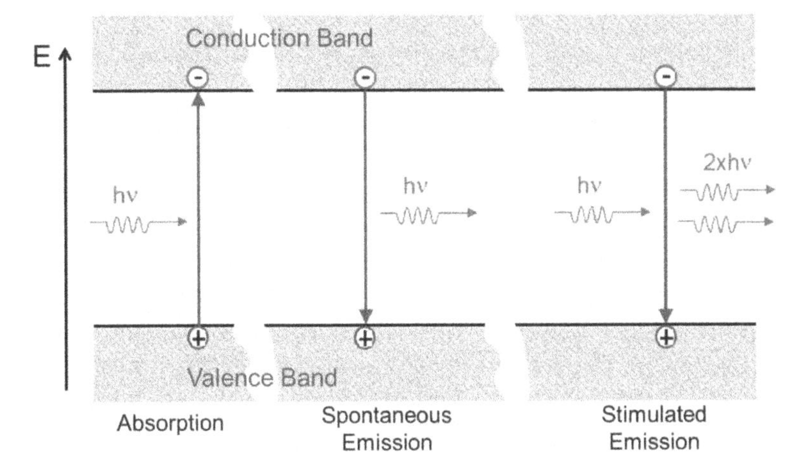

Figure 1.16 Elementary interaction processes between matter and light: Absorption (left), spontaneous emission (middle) and stimulated emission (right).

Besides the simple *absorption* of light by excitation of an electron into the conduction band and its counterpart[b] (*spontaneous emission* of light by recombination of an excited electron, see Fig. 1.16, left and middle), a third basic process is possible: The *stimulated emission* of a photon: An incoming photon of suitable energy triggers an excited electron to recombine. Then a second photon is created (see Fig. 1.16, right) which is totally identical to the incoming photon regarding its properties such as energy/wavelength, direction/momentum, and polarization. Both photons may trigger more such events while traveling through the crystal. Thus, the light intensity gets amplified; this is the key process in a laser.

For these basic processes, which can be regarded as collision of quantum particles, two laws of conservation hold:

- Conservation of energy
- Conservation of momentum (more precisely: Conservation of quantum number $\vec{k}$).

[b]Strictly speaking while taking into account quantum physics, the inverse process to absorption is *stimulated emission*, see Section 6.1.

Obviously, in these "2-particle processes", the energy of the photon $E = h\nu = \hbar\omega = \frac{hc}{\lambda}$ is hence equal to the transition energy of the electron, i.e., in the order of the band gap E_g.

The momentum of a photon is given by $\vec{p} = \hbar\vec{k} = \frac{h}{\lambda} = \frac{E}{c}$, hence $k = \frac{\omega}{c}$.

For a reasonable photon energy (in the same range as E_g, i.e., ~1 eV), the photon's momentum is very small ($<10^7\ \text{m}^{-1}$).

On the other hand, the typical momentum of the electron in a crystal may be of the order of a reciprocal lattice vector ($\frac{\pi}{a}$ (see, e.g., Fig. 1.13), i.e., $k \leq 6 \cdot 10^9\ \text{m}^{-1}$), much larger than the momentum of the photon. Therefore, the photon's momentum to be transferred to the electron is close to zero, which corresponds to a vertical transition in the $E - k$-diagram.

Thus, in a direct semiconductor, an excited electron which falls back from its excited state at the CB minimum at $k = 0$ to the VB maximum at $k = 0$ will create a photon of energy of about E_g (Fig. 1.14, left).

In an indirect semiconductor, an electron in the conduction band minimum at $k \neq 0$ cannot easily fall down to the valence band maximum at $k = 0$ (where some empty states are), because the two electron states differ not only in energy, but also significantly in momentum. The missing momentum must be put in by a third particle, typically a **phonon**, i.e., the particle related to acoustic waves (Fig. 1.14, right). Phonons can indeed have a momentum over the full range of the Brillouin zone, while their energy is still limited to few tens of milli-electron volts (see more details in solid-state physics textbooks, e.g., [1]).

The same rules hold for the inverse processes, i.e., how an incoming photon can excite an electron from the VB into the CB.[c]

Now the collision process is a three-particle process. Its probability is much less than that of the two-particle process "photon hits electron" or "electron hits photon." Therefore, the probability of radiative transitions in an indirect semiconductor is very low. This explains why Si and Ge are not used for optoelectronic

[c] If the excitation energy of the photon is large enough, then an electron may be excited at $k = 0$, i.e., into the direct minimum of the CB, which is **not** the absolute minimum in an indirect semiconductor.

devices such as light-emitting diodes.[d] Instead, an excited electron relaxes into the VB by non-radiative processes, which may be multi-phonon processes.

The recombination or relaxation is described by the time τ which tells us how long a particle stays in the excited state before it spontaneously relaxes or recombines.

The total relaxation or recombination time is then given by

$$\frac{1}{\tau} = \frac{1}{\tau_{\text{radiative}}} + \frac{1}{\tau_{\text{non-radiative}}}.$$

Most compound semiconductors have a direct band structure. Hence, they are well suited for optoelectronic applications!

1.10 Carrier Statistics

Up to now, we have seen that electronic bands are formed in semiconductors. The electronic band structure is certainly the key to understand their physical properties. Now let us discuss the still remaining basic question:

- How many carriers are actually in the conduction and valence band?

We already have argued that some electrons may be thermally excited from the originally completely filled valence band into the originally completely empty conduction band. As any such thermally controlled process, this is controlled by laws of statistics.

Let us fix the following terms by definition:

- n = concentration of electrons in conduction band
- p = concentration of holes in valence band
- **Intrinsic semiconductors** are pure materials without any impurities or doping. Thus only electrons which are (thermally)

[d]Interestingly, Si is, however, fairly well-suited for solar cells. The weak absorption of light is compensated in such devices by thicker material. The generated electrons and holes then have a fairly long life time because of the indirect band structure, simplifying their efficient separation.

excited from the valence band can be found in the conduction band. We will discuss doped semiconductors later.

We now can separate our original question into two parts:

- How many states for carriers are available in a band? In particular: How many states at a given energy? This is what we call "density of states."
- How are these states occupied at a given temperature?

An analogon of our real world may be: A lecture hall has a well-defined number of seats which can be occupied by students. This represents the available states. The students occupying these seats in a given lecture then correspond to the carriers which may occupy the states in an electronic band.

1.10.1 Density of States

In every band, there is a well-defined number of **states** which can be occupied. Under somewhat simplified assumption, this number can be found by the following calculation:

Every atom which forms the crystal delivers one state (which may be occupied by two electrons having opposite spin).

The number of these states is equal to the number of solutions of the respective Schrödinger equation for the wave functions $\Psi(\vec{r}, t)$ of the particles to be considered here, i.e., the carriers, which are characterized by their momentum $\vec{k}$.

For simplicity, we assume again that we can describe the wave functions by sinusoidal waves travelling through the whole crystal:

$$\Psi(\vec{r}, t) = \Psi_0 e^{j(\vec{k}\vec{r} - \omega t)}$$

As discussed above, periodic boundary conditions with periodicity L in all three dimensions are assumed.

Then the allowed eigenvectors which fulfill such periodic boundary conditions are

$$\vec{k} = \left(2n_x\frac{\pi}{L}, 2n_y\frac{\pi}{L}, 2n_z\frac{\pi}{L}\right) \quad \text{with}$$
$$n_i = \ldots -3, -2, -1, 0, 1, 2, 3, \ldots \text{ (integers).} \qquad (1.30)$$

Here, n runs up to $n\frac{\pi}{L} = \frac{\pi}{a}$, i.e., all these states are in the first Brillouin zone (BZ). Larger n would not lead to new solutions, as these states would be outside of the first BZ and can be folded back into the first BZ.

Equation 1.30 is an expression how the states are distributed in k-space. How are they distributed over energy?

The total number of states up to an energy E can be determined as follows:

As E increases with k^2 (see Eq. 1.10 and 1.27), the limiting energy is related to an upper magnitude $|k_E|$ of the wave vector $\vec{k}$: From Eq. 1.27 we deduce

$$k_E = \left(\frac{2m^* E}{\hbar^2}\right)^{1/2},$$

The states for all $|k| \leq |k_E|$ can be found in a sphere with radius $|k_E|$ (see Fig.1.17) and a volume

$$V_E = \frac{4}{3}\pi k_E^3.$$

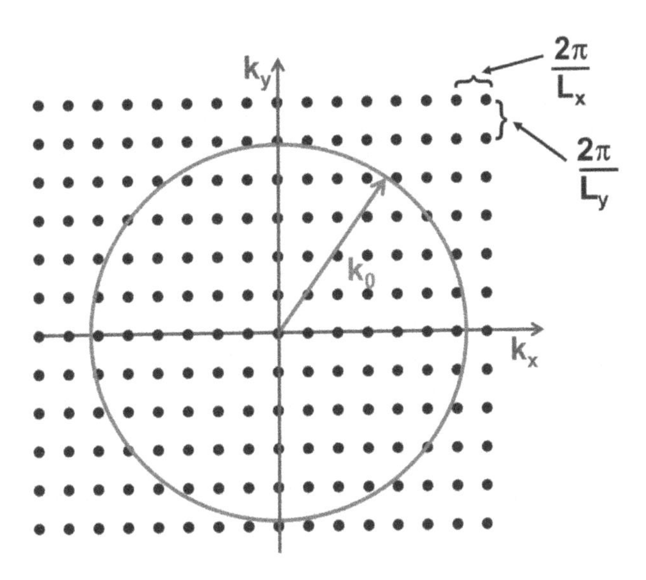

Figure 1.17 Visualization of a sphere in k-space with radius k_0 enclosing the respective reciprocal lattice points.

One state occupies a cube with length $\Delta k = \frac{2\pi}{L}$, i.e., the volume of one state is

$$V_s = \left(\frac{2\pi}{L}\right)^3.$$

Thus the number of states in our sphere is

$$N' = \frac{V_E}{V_s} = \left(\frac{L}{2\pi}\right)^3 \cdot \frac{4}{3}\pi k_E^3 = \frac{V}{6\pi^2}\left(\frac{2m^* E}{\hbar^2}\right)^{3/2}$$

within our considered macroscopic volume $V = L^3$.

Owing to the two spins of the electrons, any such state represents in fact two states. Thus we get the number of states **per volume**:

$$N = 2 \cdot \frac{N'}{V} = \frac{1}{3\pi^2}\left(\frac{2m^* E}{\hbar^2}\right)^{3/2}$$

Hence the density of states $D(E)$ reads

$$D(E) = \frac{dN}{dE} = \frac{1}{2\pi^2}\left(\frac{2m^*}{\hbar^2}\right)^{3/2} \cdot E^{1/2}. \quad (1.31)$$

In the conduction band, the energy zero point is at $E = E_c$. Moreover, in each band, the effective mass may be different (m_e^* for electrons in the CB, m_h^* for holes in the VB) . Thus we get for the conduction band density of states:

$$D_c(E)dE = \frac{1}{2\pi^2}\left(\frac{2m_e^*}{\hbar^2}\right)^{3/2} \cdot (E - E_c)^{1/2} dE \quad (1.32)$$

and for the holes in the valence band:

$$D_v(E)dE = \frac{1}{2\pi^2}\left(\frac{2m_h^*}{\hbar^2}\right)^{3/2} \cdot (E_v - E)^{1/2} dE \quad (1.33)$$

These densities of states describe how many states we may find in the energy interval between E and $E + dE$ (Fig. 1.18, left).

Note that for a 3D crystal,

$$\boxed{D(E) \sim \sqrt{E}}$$

Later, we will learn how this relation can be changed in structures with lower dimensions (see Chapters 8 and 10). This is a very significant change concerning the functionality of some optoelectronic devices.

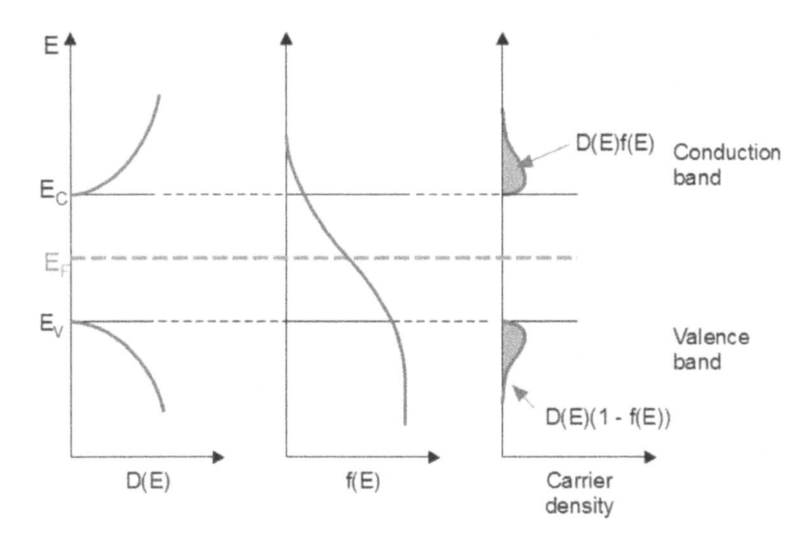

Figure 1.18 Density of states (left), Fermi distribution (middle) and carrier density distribution (right) in an intrinsic semiconductor (schematically).

1.10.2 Fermi Distribution

Now the question that remains to be answered is: How are these states occupied?

The probability whether an available state is occupied is governed by the Boltzmann probability:

$$p(E) \sim e^{-\frac{E}{kT}}$$

Remember: Electrons are Fermi particles, i.e., they are governed by the Pauli principle: Once a state is occupied, it cannot be occupied by another particle with exactly the same quantum numbers.

Thus the electrons are distributed over the available states according to the Fermi distribution probability:

$$f(E) = \frac{1}{1 + e^{\frac{E - E_F}{kT}}}, \tag{1.34}$$

where E_F is the Fermi energy (see Fig. 1.19). At $T = 0$ all states up to E_F are occupied, all others are empty. Hence, the Fermi energy E_F describes the situation of thermodynamic equilibrium. It is exactly the same as the chemical potential known from chemical reactions.

Now, we can calculate the number of carriers in the conduction band at $T > 0$ by multiplying the density of states $D(E)$ with the probability of their occupation $f(E)$, then integrating over all energies (see Fig. 1.18, right):

$$n(T) = \int_0^\infty D(E)f(E)dE$$
$$= \int_{E_c}^\infty \frac{1}{2\pi^2}\left(\frac{2m_e^*}{\hbar^2}\right)^{3/2} \cdot (E - E_c)^{1/2} \frac{1}{1+e^{\frac{E-E_F}{kT}}} dE \quad (1.35)$$

which finally results in

$$n(T) = 2\left(\frac{m_e^* kT}{2\pi\hbar^2}\right)^{3/2} \cdot e^{\frac{E_F-E_C}{kT}} = N_C \cdot e^{\frac{E_F-E_C}{kT}} \quad (1.36)$$

with the so-called **effective density of states** of the conduction band

$$N_C = 2\left(\frac{m_e^* kT}{2\pi\hbar^2}\right)^{3/2}. \quad (1.37)$$

Similarly, we obtain for the hole states in the valence band:

$$p(T) = 2\left(\frac{m_h^* kT}{2\pi\hbar^2}\right)^{3/2} \cdot e^{\frac{E_V-E_F}{kT}} = N_V \cdot e^{\frac{E_V-E_F}{kT}}, \quad (1.38)$$

with N_V being the effective density of states of the valence band.

Please notice: By increasing E_F we can increase n by simultaneously decreasing p! This law holds also in the inverse sense:

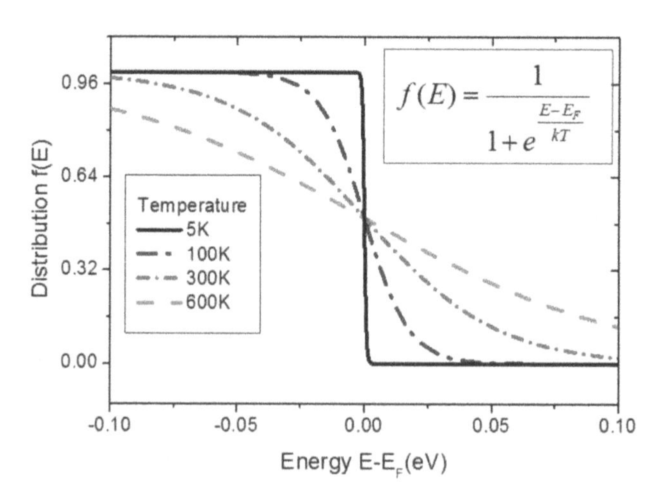

Figure 1.19 Fermi distribution function for different temperatures.

By changing the carrier concentration in a semiconductor, we can change the position of the Fermi level E_F.

In an intrinsic crystal, the number of thermally excited electrons in the CB $n_i(T)$ is exactly the same as the number of holes in the VB $p(T)$. This enables to determine the position of the Fermi energy for this case:

$$E_\mathrm{F} = \frac{E_c + E_v}{2} - \frac{3}{4}kT \ln \frac{m_e^*}{m_h^*} \tag{1.39}$$

which can be simplified to

$$E_\mathrm{F} = \frac{E_c + E_v}{2} \tag{1.40}$$

if the effective masses are about the same. This means the Fermi energy lies half-way between the valence band (completely filled at $T = 0$) and the conduction band (completely empty at $T = 0$) (see Fig. 1.18, middle).

Now, the carrier concentration in the conduction band can be described as if all carriers occupy the band edge, governed by the Boltzmann factor:

$$n(T) = N_C e^{-\frac{E_g}{2kT}} \tag{1.41}$$

We see that the carrier concentration increases drastically with T, hence the conductivity increases also with T, as already mentioned as one of the basic properties of a semiconductor (for some details see later).

From equations 1.36 and 1.38, we can deduce the so-called **mass action law**:

$$n(T) \cdot p(T) = n_i^2 = 4\left(\frac{kT}{2\pi\hbar^2}\right)^3 \cdot (m_e^* m_h^*)^{3/2} e^{-\frac{E_g}{kT}}, \tag{1.42}$$

i.e., the product $n \cdot p$ does not depend on E_F! This law is also valid for non-intrinsic semiconductors (in thermal equilibrium), particularly for doped semiconductors (see Section 5.1). It means that if we increase the electron concentration $n(T)$ in the CB, then $p(T)$ in the VB decreases accordingly.

1.11 Defects in Semiconductors

Here, we like to only briefly classify some typical defects which can be found in compound semiconductors and are relevant for devices made of these materials. Most of these defects are also present in other crystals or solid state materials.

A common classification can be done by considering the dimensionality of the defect:

(1) Zero-dimensional defects, point defects:
As the name suggests, those are defects typically related to single atoms. The following types are common:

- A missing atom in a crystal is called a **vacancy**. This may be caused by non-stoichiometric crystal growth in compound semiconductors. A typical example is missing N atoms in GaN owing to the large chemical stability of NH_3 and thus reduced supply of atomic nitrogen during crystal growth (see Chapters 3 and 4).
- If a "wrong" atom occupies a lattice site, e.g., a dopant such as Se occupying an As site in GaAs, then this is called a **substitutional** defect. Such doping is used to intentionally change the electrical properties of a semiconductor (see Chapter 5). In many cases, such impurities are incorporated unintentionally. A special case is that a regular atom of a crystal occupies the opposite site, e.g., Ga on an As site in GaAs. Then this defect is called an **antisite** defect.
- An atom positioned between the regular atoms is called **interstitial** defect. Again, this may be an atom of the host material or a foreign impurity.

(2) One-dimensional defects, line defects:
Typical line defects are **dislocations**, a very significant class of crystalline defects. They are caused by a missing or shifted line of lattice points. The typical examples are (see Fig. 1.20):

- edge type dislocation, where a row of atoms is missing (see Fig. 1.20 left);
- screw type dislocations which can be explained as follows: A perfect crystal may be cut along a plane, then one half

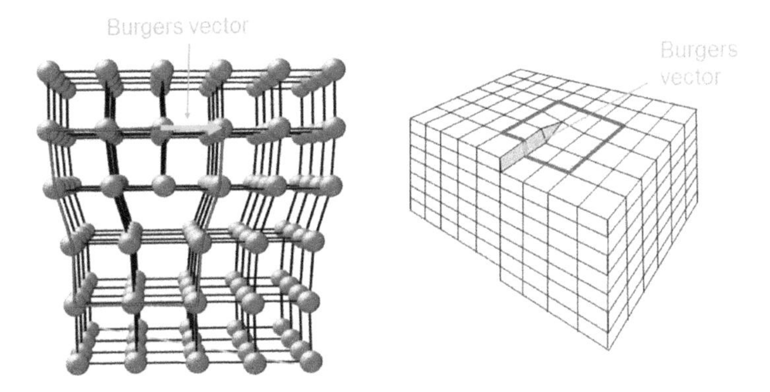

Figure 1.20 Edge type (left) and screw type dislocation (right).

is slipped across the other by a lattice vector, before fitting them back together without leaving a defect (see Fig. 1.20 right).

In many cases, mixed dislocations form having properties of both types. Moreover, partial dislocations are quite common in crystals containing more than one atom in the basis.

A dislocation can be nicely described by the so-called **Burgers vector** $\vec{b}$. This vector can be found as follows: In a lattice, move a number m of steps along one crystal axis x, then another number n along a perpendicular axis y. From the final point, move back in $-x$-direction by the same amount of steps m and then in $-y$-direction by n steps. If you do not end at the origin of your path, then a dislocation is enclosed in your path. The vector which connects the end point and the starting point of this open loop is the Burgers vector (see Fig. 1.20). This vector is oriented perpendicular to the dislocation line for edge dislocations, whereas it is parallel for screw dislocations.

Dislocations typically are responsible for non-radiative optical transitions and therefore are harmful for many optical devices decreasing their radiative efficiency (see Section 12.1.2). Moreover, the carrier mobility may be limited by them, and even the life time of laser diodes is strongly affected by them, leading to the need of extremely low dislocation densities in GaAs wafers (see Chapter 3).

The density of dislocations is quantified by measuring the length of all dislocation lines in a given volume, i.e., it is measured in units of $[\text{cm}/\text{cm}^3] = [\text{cm}^{-2}]$.

If all dislocations run parallel in one direction, e.g., as so-called "threading dislocations" from the substrate to the surface of the epitaxial layer,[e] then the density of the threading points on the surface reflect the same quantity (measured in units of dots per cm^2). Such numbers are also taken to characterize, e.g., wafers (see Chapter 3).

(3) Two-dimensional defects, areal defects:

Typical areal defects in compound semiconductors are related to the stacking sequence (see Section 1.3). They are particularly important in non- and semipolar GaN.[f] The wurtzitic lattice of GaN contains the layer sequence A-B-A-B-A-B ..., whereas the cubic lattice of GaAs, etc., follows the sequence A-B-C-A-B-C-A-B-C ... A **stacking fault** is an areal defect, where the regular sequence is disturbed. Typical examples in wurtzitic crystals are the sequences

- A-B-A-B-**C**-B-C-B-C ... (Type I_I),
- A-B-A-B-**C**-A-C-A-C ... (Type I_{II}),
- A-B-A-**C**-A-B-A-B ... (Type I_{III}), and
- A-B-A-B-**C**-A-B-A-B ... (extrinsic).

Hence they can be regarded as cubic inclusions into the hexagonal lattice.

Other popular areal defects in compound semiconductors are **twins** describing situations, where two crystals of the same kind but different orientation hit at an interface which is still perfect with respect to all bonds. A simple example is the sequence A-B-C-A-B-C-A-B-C=B-A-C-B-A-C-B-A ... (twin boundary marked by "="), i.e., a cubic twinned crystal. Such twins may occur in bulk crystal growth as of course both parts of the twinned crystal have the same probability to develop.

These defects are quantified by measuring the total area of the defects in a given volume, i.e., this is measured in units of

---

[e] Typical for GaN grown on foreign substrates, see Chapter 11.

[f] Less polar GaN structures obtained by growing the epitaxial structure not in the conventional c-direction as briefly mentioned in Section 11.6.

$[\text{cm}^2/\text{cm}^3] = [\text{cm}^{-1}]$. Again: If the areal defects run in parallel and penetrate the surface, then their penetration line may be made visible. Hence their density of total length of penetrating lines per given area reflects also the stacking fault density.

(4) Three-dimensional defects: For completeness, three-dimensional defects like grains, inclusions, voids, etc., are also briefly mentioned, although they are typically easily avoided and hence do not play a significant role in our materials.

Problems

(1) (a) Give some arguments which show that crystals may have electronic bands.
    (b) Explain the difference between metals, semiconductors, and insulators.
    (c) How does the electrical conductivity change with temperature in an (intrinsic) semiconductor? Why?

(2) (a) Explain: What is the Wigner–Seitz cell of a crystal?
    (b) What requirements must this cell fulfill with respect to translation symmetry?

(3) Why is Si not a good material for light-emitting diodes? Which basic laws must be fulfilled, which may be difficult in Si?

(4) Explain the term "first Brillouin zone" and its importance for understanding semiconductors.

(5) (a) In which situation can we find a well-defined Fermi level? What does it represent?
    (b) Why is the Fermi level even at $T = 0$ K at a finite energy?
    (c) Calculate the position of the Fermi level of intrinsic GaAs at room temperature. Hint: For the valence band, take only the heavy holes into consideration. You may find the necessary material properties in the script or at http://www.ioffe.rssi.ru/SVA/NSM/Semicond/index.html.
    Make use of the fact that in an intrinsic semiconductor, the concentrations of electrons in the conduction band and the holes in the valence band are the same.

(6) (a) Where is the Fermi level in moderately n-doped GaAs (at room temperature)?

(b) How does the carrier concentration change with temperature in such a semiconductor (just qualitatively)?
(c) So how does the position of the Fermi level change with temperature?

(7) (a) Sketch how to calculate the density of states of a semiconductor.
(b) Does it change with temperature?
(c) How does it change with energy? Is this good or bad?

(8) What is a dislocation? How can it be described fairly quantitatively?

(9) What is a stacking fault?

Chapter 2

Introduction to Compound Semiconductors

In the last chapter, we discussed some basics of semiconductors in general. The best known and technically still most important semiconductor is silicon (Si). Like germanium (Ge), it is an elemental semiconductor. Both have a band gap around 1 eV (see Table 2.1). These elements are positioned in the center of the periodic system of elements (PSE). They have four valence electrons forming four covalent bonds in tetragonal symmetry resulting in a diamond-like crystal structure (fcc). Indeed, diamond is another representative of this class of semiconductors, having a fairly large band gap of $E_g \simeq 5.5$ eV. These materials have an *indirect* band structure.

In this book, we focus on semiconductors, which are formed by more than just one sort of elements, the so-called **compound semiconductors**. They are formed by elements positioned around those elemental semiconductors in the PSE.

The most typical example is **GaAs**, formed by the following:

- **Ga**: In the PSE it is the left neighbor of Ge. It has three valence electrons.
- **As**: It is the right neighbor of Ge with five valence electrons.

Compound Semiconductors: Physics, Technology, and Device Concepts
Ferdinand Scholz

ISBN 978-981-4774-07-9 (Hardcover), 978-1-315-22931-7 (eBook)
www.panstanford.com

Table 2.1 Characteristic data of some compound semiconductors at room temperature, compared to those of the elemental semiconductors Si and Ge

Compound	Lattice constant (nm)	Band gap (eV)	Carrier mobility (cm^2/Vs)
Ge	0.5658	0.661	3900
Si	0.5431	1.12	1500
InSb	0.6479	0.17	100,000
InP	0.58687	1.344	5400
GaAs	0.5653	1.42	8500

In total, the two atoms contribute eight valence electrons as would do two Si atoms. Therefore, GaAs crystallizes in basically the same structure as C, Si and Ge: zinc blende instead of diamond, because the two atoms of the basis are now different (c.f. Section 1.3).

Many other features are not so different to Si and Ge. Some exemplary data are given in Table 2.1. However, in strong contrast to the elemental semiconductors, most (although not all) compound semiconductors, including GaAs, have a direct band structure (Fig. 2.1).

In fact, there is a big choice of forming such compound semiconductors by combining elements of group III of PSE with those of group V (→ III-V compounds such as GaAs, InP, GaN, GaSb, and AlP; see Fig. 2.3) or even of group II and group VI (→ II-VI compounds such as ZnSe, MgO, and CdS).

Indeed, these various materials differ significantly in many properties, not only in the band gap and the lattice constant. This opens excellent possibilities to choose the appropriate compound for the required application. Moreover, they may have other attractive features like small effective electron masses resulting in large carrier mobilities (see Table 2.1).

Besides the binaries, there is the chance of forming ternary, quaternary (and even more complex) alloys like $Al_xGa_{1-x}As$, $Ga_xIn_{1-x}As$, $Ga_xIn_{1-x}As_yP_{1-y}$, etc.

Typically, the material properties of ternary or higher alloys can be extracted from those of the binaries by a linear interpolation described by **Vegard's law**:

$$a_{Ga_xIn_{1-x}As} = x \cdot a_{GaAs} + (1 - x) \cdot a_{InAs}, \tag{2.1}$$

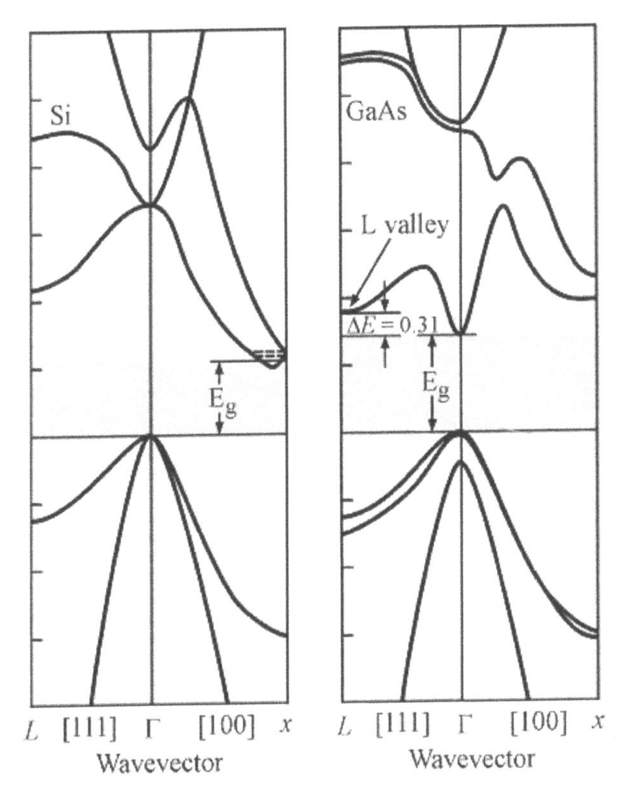

Figure 2.1 Indirect band structure of Si (left) and direct band structure of GaAs (right) (after [4]).

where a describes some property of the compound. This is in particular true for the lattice constant.

Some other features, in particular the band gap, may require a quadratic relation:

$$E_{g,\mathrm{Ga_xIn_{1-x}As}} = x \cdot E_{g,\mathrm{GaAs}} + (1-x) \cdot E_{g,\mathrm{InAs}} - x \cdot (1-x) \cdot b, \quad (2.2)$$

where b is called the **bowing parameter**.

This means that most material properties (lattice constant, band gap, refractive index, thermal properties, mechanical constants etc.) can be adjusted smoothly by the composition of these alloys. This increases further significantly the freedom of choosing the best compound for a particular application.

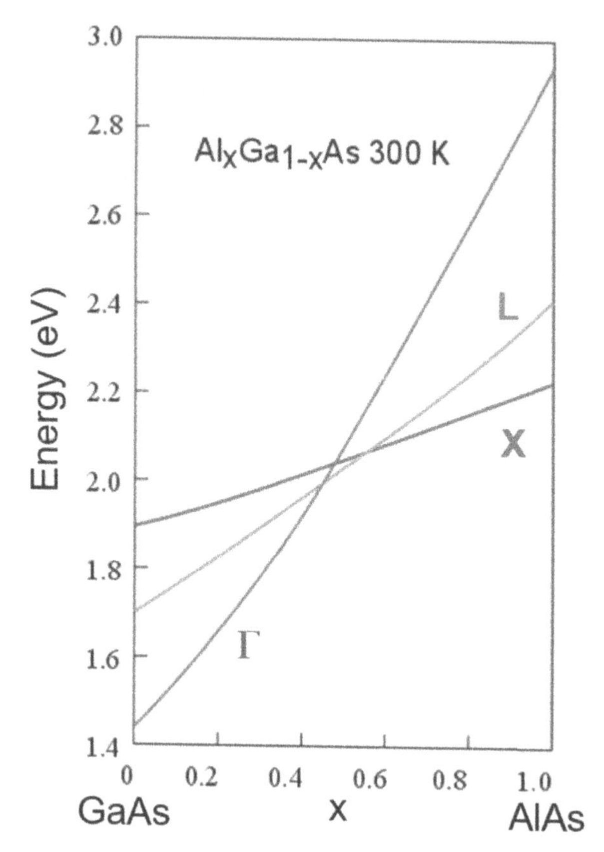

Figure 2.2 Transition from direct to indirect band structure when changing the composition *x* of $Al_xGa_{1-x}As$ (data from [5]).

Please notice: Different electronic bands may follow different equations. Example: GaAs is a direct semiconductor having the absolute conduction band minimum at $k = 0$ (Γ point). Other conduction band minima are at $k \neq 0$ in different directions (e.g., X valley in <100> direction and L valley in <111> direction). On the other hand, AlAs is an indirect compound semiconductor where the absolute conduction band minimum is the X valley. Therefore, $Al_xGa_{1-x}As$ makes a transition from being a direct semiconductor for $x < 0.45$ to an indirect semiconductor for $x > 0.45$ (see Fig. 2.2), as the parameters in the band gap equation 2.2 are different for each minimum.

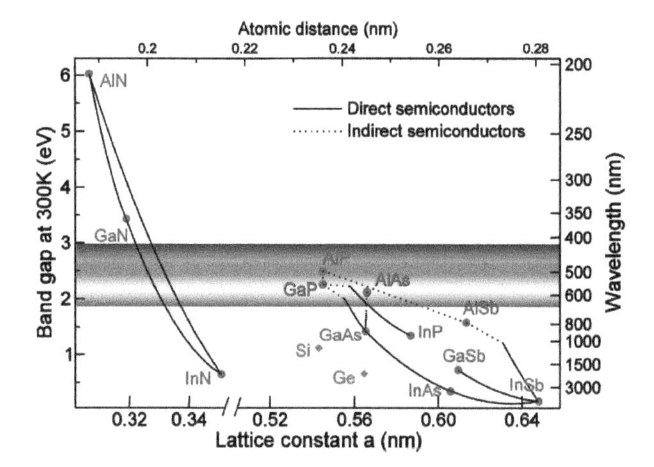

Figure 2.3 Band gap versus lattice constant of some compound semiconductors. The lines represent some ternary systems. The top x-axis shows the direct atomic distance between the two atoms forming the basis of these crystals. The right axis shows the wavelength corresponding to the band gap, the visible range is indicated by the rainbow colors.

However, the synthesis of compound semiconductors is more difficult than that of elemental semiconductors. Therefore, extra chapters are included in this book discussing these challenges. It is more or less not feasible to grow ternary bulk crystals, because the control of a uniform composition is very difficult. Some binaries such as GaAs, InP, GaSb, and GaP can be grown as bulk crystals (see Chapter 3), whereas the bulk growth of other technically very important ones like GaN is extremely challenging and still a hot topic in basic research.

Full use of the flexible properties of these compound semiconductors can be made by growing only thin layers (some micrometers thick or even thinner), as discussed in Chapter 4.

There may be limitations in particular for the synthesis of ternary and quaternary compounds, where miscibility gaps exist, e.g., in the center of the system $Ga_xIn_{1-x}As_yP_{1-y}$ (Fig. 4.5). Other compounds like $Al_xGa_{1-x}As$ are miscible in the full composition range.

Growth of thin layers by epitaxy needs growth on some kind of substrate. Therefore some boundary conditions should be (more or less) fulfilled:

- Substrate and layer should have same crystalline structure.
- They should not differ too much in lattice constant.

There are simple systems, in particular $Al_xGa_{1-x}As$, because the lattice constants of GaAs and AlAs are very close:

$$\frac{a_{\text{GaAs}} - a_{\text{AlAs}}}{a_{\text{GaAs}}} = \frac{\Delta a}{a} \approx 1.2 \cdot 10^{-3}$$

Here, the lattice matching condition is fulfilled for effectively every composition x.

However, this is not the case for $Ga_xIn_{1-x}As$, because here

$$\frac{a_{\text{InAs}} - a_{\text{GaAs}}}{a_{\text{GaAs}}} = \frac{\Delta a}{a} \approx 7\,\%.$$

Hence, only a well-defined composition may be regarded to be lattice matched to a given substrate (which may be InP with a lattice constant just between GaAs and InAs).

If the lattice matching condition is fulfilled, there is still a big choice of binary, ternary and quaternary materials to be grown on a given substrate opening the field to synthesize complex multi layer **heterostructures**. Hence, this can be used to design complex device structures which will be a major focus in this book. Sometimes these possibilities are called **band gap** or **band structure engineering**. We will learn even about many other chances how to synthesize materials and structures with totally novel properties.

If two different semiconductor layers are grown epitaxially one on top of each other thus building a heterostructure, then they are connected at the **heterojunction**. At this interface, the band structure of both materials must be connected in some way (Fig. 2.4). Here, the vacuum level is important, i.e., the energy (electron affinity χ) which is needed to move an electron from the conduction band edge into the vacuum. A related figure is the work function ϕ which is the energy needed to lift an electron from the Fermi level to the vacuum. According to Anderson's rule, the vacuum levels of the two semiconductors on either side of the heterojunction should be aligned smoothly at the same energy. Accordingly, band offsets ΔE_C and ΔE_V form at the conduction band and valence band connection (see Fig. 2.4), the size of which is given by the combined material properties. When connected, the bands in distance to the junction line up as governed by the Fermi level, still

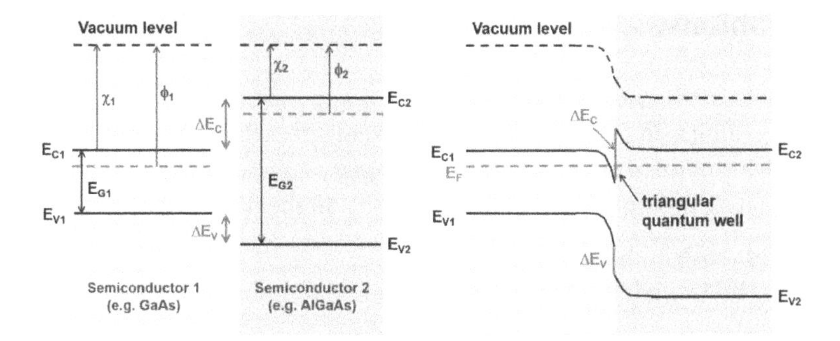

Figure 2.4 Band offsets at a heterojunction: Left: unconnected. The bands are lined up with reference to the vacuum levels governed by the electron affinity χ_i thus defining the offsets ΔE_C and ΔE_V in the conduction and valence bands, resp. Right: When connected, these band offsets remain at the interface. According to the bulk position of the Fermi level, some band bending at the interface will develop. Here, we assumed n-type materials on either side of the junction.

obeying Anderson's rule at the interface. This may result in some band bending at the interface. Depending on the positions of the Fermi levels, a triangular quantum well at the interface may form (see Fig. 2.4 right, more details: see Chapters 8 and 13).

Depending on the size and signature of the band offsets, different types of interfaces can form (Fig. 2.5). In most cases, a type I heterojunction is formed. We will mainly discuss heterostructures with this kind of interface, which directly enable most technically relevant devices such as light-emitting diodes (LEDs), laser diodes, field-effect transistors, etc.

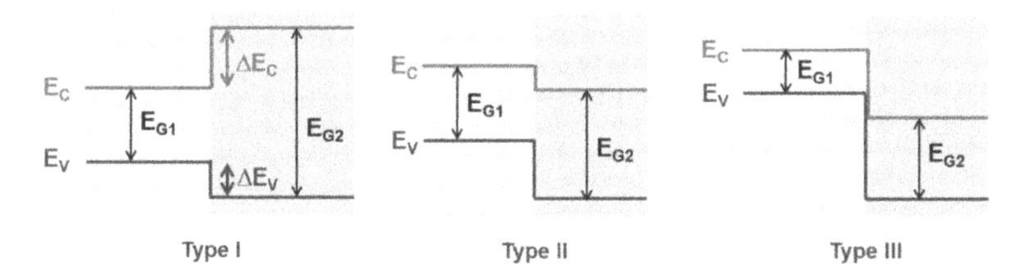

Figure 2.5 Different band alignments at heterojunctions. In most cases, type I interfaces form. A type II alignment can be found at an AlInAs-InP interface, whereas a type III alignment forms, e.g., at the interface between InAs and GaSb.

Problems

(1) Describe some advantages of III-V compound semiconductors as compared to the elemental semiconductors Si and Ge.

(2) From the data of the respective binary compound semiconductors, determine the lattice constant and the band gap of

  (a) $Al_{0.3}Ga_{0.7}As$
  (b) $Ga_{0.47}In_{0.53}As$

  Determine the lattice mismatch between both compounds and the respective binaries. How large is the lattice mismatch of $Ga_{0.47}In_{0.53}As$ with respect to InP?
  The necessary data can be found at Web link indicated in problem 5 of Chapter 1.

(3) What makes the material system AlGaAs particularly interesting for many device applications?

(4) GaAs has a direct band structure, whereas AlAs has an indirect structure. How about $Al_xGa_{1-x}As$? Explain.

(5) Which properties of the constituents of a heterostructure should you know to determine the band offsets in the conduction and valence band, if you connect them? Explain.

Chapter 3

Bulk Crystal Growth

In this book, we provide a very short overview of the field of bulk growth of compound semiconductor crystals. The interested reader may find more details in [6, 7].

As already mentioned, compound semiconductors find their main applications in many optoelectronic and electronic devices. The functionality of most of these devices is given by the design of a specific heterostructure. This means, in consequence, that devices are typically not made of bulk crystals, but by epitaxially grown multi-layer structures (see Chapter 4).

Anyway, bulk crystalline material is an important prerequisite for such devices, as such heterostructures need to be deposited on a well-suited substrate. In many cases, so-called wafers are used as substrates, which are cut out of bulk material of the same kind as the device structure grown on it (see below).

Consequently, the quality of such wafers may significantly determine some properties of the final device. Besides their basic characteristics like lattice constant or doping, their crystalline quality plays a significant role.

On the other hand, the growth of crystals containing more than one element is much more challenging than single-element material

Compound Semiconductors: Physics, Technology, and Device Concepts
Ferdinand Scholz

ISBN 978-981-4774-07-9 (Hardcover), 978-1-315-22931-7 (eBook)
www.panstanford.com

Table 3.1 Melting point (MP) and vapor pressure of group V constituent at MP of some binaries

Compound	Melting Point (°C)	Vapor Pressure (10^5 Pa)
InSb	525	4×10^{-8}
InAs	943	0.33
InP	1062	27.5
GaSb	712	10^{-6}
GaAs	1238	0.976
GaP	1465	32
GaN	2500	3×10^{4}
Si	1410	5×10^{-8} (@ 1300°C)

(such as Si and Ge). Therefore, strong efforts have been made over the recent years to optimize bulk semiconductor materials.

With respect to the crystalline growth of bulk compound semiconductor materials, the following properties are of major importance (see Table 3.1):

- melting point of the semiconductor material
- vapor pressure of group V constituent, as this typically determines the stability of such materials at high temperatures

If this vapor pressure is small enough (e.g., for antimonides), then no specific stability problems arise at temperatures around the melting point. Consequently, similar methods as well established for elemental semiconductors, e.g., Si and Ge, can be applied.

However, for many III-V compounds, the equilibrium vapor pressure of the group V element over the solid material is large, sometimes significantly above ambient pressure. Then, crystalline growth at "normal" conditions is not possible.

In order to suppress the uncontrolled evaporation of the group V constituents, mainly two methods have been developed:

- Hot wall technology: In a closed environment, keep the walls of the reaction chamber hot so that the group V material does not condensate, but remains in the gas phase in equilibrium with the growing crystal.

- Liquid encapsulation technology, i.e., sealed coverage of the melt in order to suppress any evaporation of group V material out of the melt.

We will discuss these points briefly mainly along GaAs as a representative example. A fairly recent review paper about bulk growth of GaAs was published by Rudolph et al. [8] focusing on some respective details.

3.1 Czochralski Method

The **Czochralski** crystal growth method was heavily optimized for the growth of Si single crystals which today form the basis of all our modern electronics. Its working principle is as follows: The material to be crystallized is molten, the melt has a temperature close to the melting point: $T \geq T_{MP}$. Then, a seed crystal is immersed, cooling the melt locally, hence the crystal starts growing. The seed is rotated and slowly pulled out of the melt. Hence, the growth of a single crystal propagates from the seed.

The following preconditions should be fulfilled, if applied to compound semiconductors:

- Congruent or nearly congruent melting (transition of a multi-component system from solid to liquid phase without change of concentration).
- No destructive phase transition during cooling down from solidification to room temperature
- Vapor pressure should not be extremely high at melting point. Otherwise (e.g., GaAs and InP): Liquid encapsulation (see below).

Liquid encapsulated Czochralski method (LEC) (Fig. 3.1): In order to prevent the evaporation of components with high equilibrium vapor pressure, the melt can be covered. For the best sealing at the outer walls of the crucible, etc., this task is best fulfilled by an adequate liquid floating on the melt. Hence, the liquid encapsulant should have the following properties:

- Not miscible with melt;

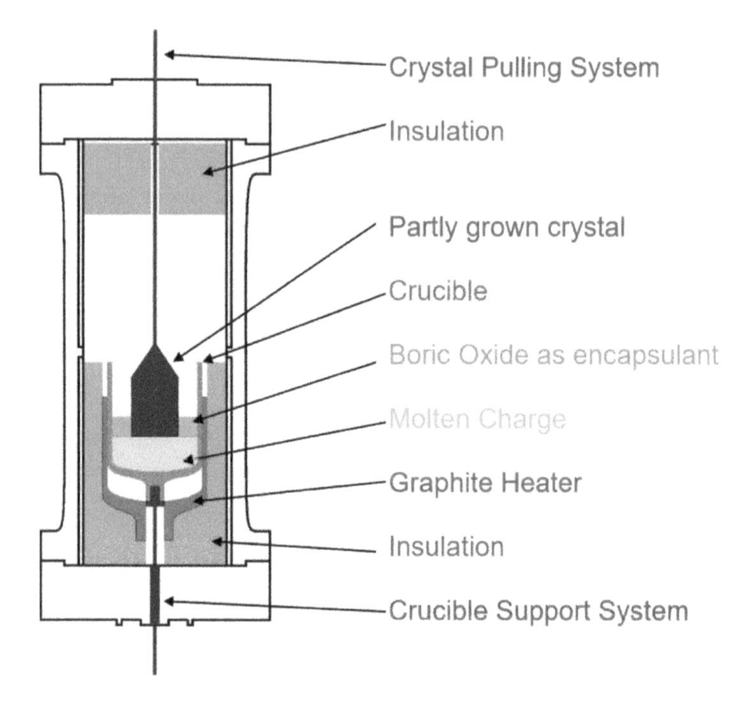

Figure 3.1 Liquid encapsulated Czochralski (LEC) method (schematically, after [9]).

- Chemically inert against melt and crucible;
- Must wet melt and crucible;
- Viscosity and temperature dependence of viscosity should be so that even the pulled-out crystal is still covered with a thin layer;
- High purity, should not release any impurities into the melt;
- Should float on the melt;
- Low vapor pressure.

A typical well-suited liquid encapsulant material is B_2O_3.

The crucible must be stable at high temperatures, also with respect to chemical interactions. Typical crucible materials are

- quartz;
- glassy carbon;
- AlN;

- SiN;
- pyrolytic BN.

For the growth of GaAs, a fairly simple machinery can be used, because only low-pressure LEC (max. $2\text{–}3 \times 10^5$ Pa) is needed. However, in industrial production, higher pressures (e.g., 2 MPa) are used [8]. Even higher pressures (3–4 MPa) are needed for other materials such as InP and GaP.

The main advantages of the LEC method are as follows:

- There is free growth and no contact of solid materials to crucible walls.
- Crystals with circular shape can be easily grown.
- B_2O_3 can getter impurities.
- Preconditioning of the melt is possible (e.g., for doping).
- The high counter pressure can be realized with some inert gas, not necessarily with the constituent (e.g., As and P) itself.

However, this method also has some inherent disadvantages:

- Control of stoichiometry is difficult, as some evaporation through the B_2O_3 encapsulating layer may take place.
- Evaporation of As (in case of GaAs) from the crystal surface may occur.
- At the liquid–solid interface, large temperature gradients (100–150 K/cm) develop, which lead to thermal strain and in consequence to the formation of dislocations.
- Thermal convection in the melt leads to uncontrolled fluctuations concerning doping or strain.
- The machinery is fairly expensive, and high investment is required.
- Post-growth multi-step heating is necessary in order to improve the residual stress level and the homogeneity of the electrical properties.

For the best crystals, a very good control of the process is mandatory. First of all, an excellent control of the pulling speed is required to obtain a crystal with uniform diameter. This may be done by continuous weighting or by optical or x-ray observation methods. It should be mentioned that such in situ methods can be implemented into the LEC process fairly easily.

Convections in the melt can be minimized by, e.g., magnetic fields thus reducing temperature fluctuations from 18 K down to 0.1 K!

One basic problem of Czochralski, in particular when growing InP, is the **formation of twin defects**. These faults have a very small formation energy. Twins could be suppressed by a larger axial temperature gradient, but then dislocations grow. For larger diameters (above 75 mm) gradients below 2–3 K/cm would be necessary to suppress dislocations, but then decomposition and twin formation problems increase.

Another chance: **Impurity hardening**, e.g., In doping of GaAs or S doping of InP for suppression of dislocations.

In the past, LEC was the most important method for the growth of GaAs bulk material. Typical industrial data are as follows:

- Ingot diameter: 200 mm
- Growth speed: 7 mm/h
- Ingot length for such diameter: 60 mm and more

However, the wafers cut out of these crystals have a high dislocation density of several 1000 cm^{-2}. This may be fine for electronic devices, but not acceptable for optoelectronic devices, in particular laser diodes.

3.2 Bridgman Crystal Growth

Another growth method for single crystals is the **Bridgman** growth method. It works basically as follows: A closed crucible filled with the raw material is moved through a hot zone where the material melts. On the side leaving the hot zone, crystal growth proceeds. A major advantage as compared to the Czochralski method is the significantly smaller thermal strain at the liquid–solid interface. Therefore, lower dislocation densities can be obtained.

Originally, this was applied to horizontally moving crucibles. Then the method is called **Horizontal Bridgman** (HB) or horizontal gradient freeze method with the following advantages:

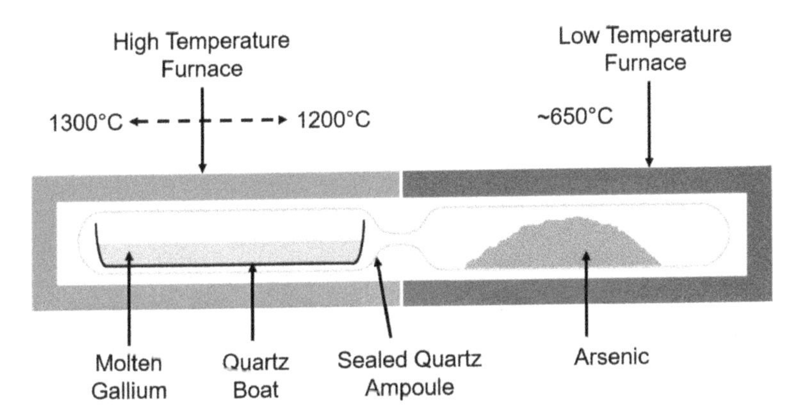

Figure 3.2 Horizontal gradient freeze method (schematically, after [9]).

- The crucible is open on the top side.
- The process can be visually controlled.

If the vapor pressure of the compound is larger than atmospheric pressure, then the process is done in a closed ampoule with additional elemental group V material (e.g., As for GaAs), the vapor pressure of which can be controlled by the temperature (Fig. 3.2).

Unfortunately, owing to the crucible shape and the shape of the liquid top surface, only wafers with D-shape can be grown.

In order to enable the growth of circular wafers, the **vertical gradient freeze** method (Fig. 3.3) was developed: Chunks of polycrystalline GaAs, produced by horizontal synthesis, are placed in a crucible with a seed crystal of the required orientation. The crucible is then placed vertically in a furnace and a temperature gradient is moved up the length of the crystal (away from the seed).

Single-crystal growth propagates from the seed crystal and, because the crystal forms in the shape of the crucible, diameter control of the ingot is relatively simple. The liquid–solid interface is horizontal and plane.

Now, the main problem is the strain as a consequence of the contact between the semiconductor and the crucible walls. This could be solved by using a pyrolytic BN crucible with an B_2O_3 intermediate layer.

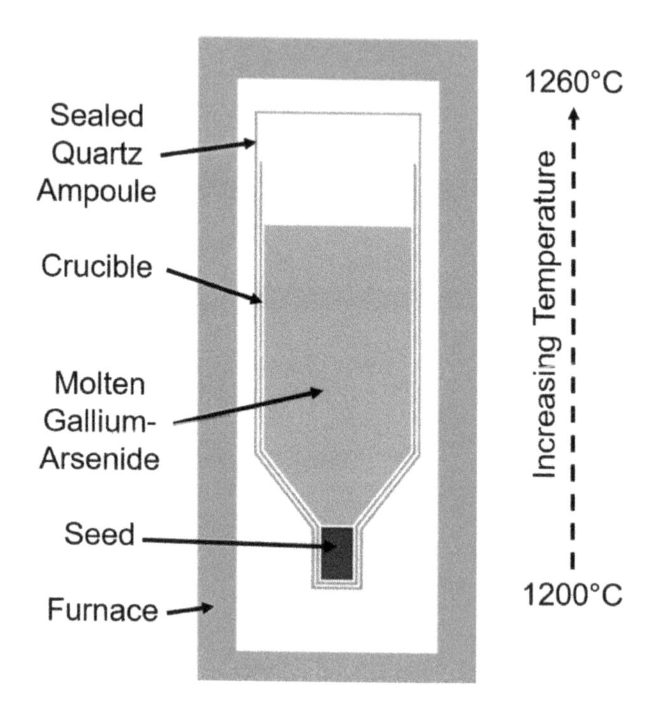

Figure 3.3 Vertical Gradient Freeze method (schematically, after [9]).

Consequently, the advantages of VGF can be summarized as follows:

- ability to fabricate circular boules and wafers
- only small temperature gradients at the liquid–solid interface and in the grown material
- uniaxial heat and material transport
- control of stoichiometry possible by separately heated As source
- low investment costs

However, there are also some significant disadvantages of VGF:

- no visual control of growth possible
- control of carbon doping difficult
- growth of long ingots difficult

Indeed, this method enables the growth of low-dislocation-density GaAs for laser diodes, etc.

State of the art (GaAs VGF):

- ingot diameter: 200 mm
- growth speed: 2–4 mm/h
- ingot length for such diameter: 140 mm
- ingot mass $\simeq$ 30 kg
- temperature gradient $\simeq 3 - 5$ K/cm
- dislocation density far below 500 cm^{-2} approaching zero

In the recent years, VGF has seen tremendous development and eventually surpassed LEC in terms of sold GaAs wafers.

3.3 Wafer Preparation

As mentioned above, the main application for such ingots is the preparation of wafers (Fig. 3.4) for subsequent epitaxial processes. Hence the main fabrication steps for such wafers (with GaAs as example) can be summarized as follows:

- GaAs high-pressure synthesis for getting the raw material
- crystal growth
    - boule growth as described above (VGF or LEC)

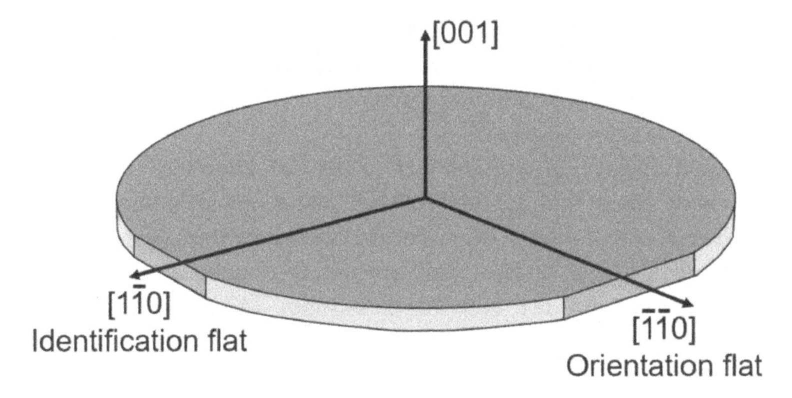

Figure 3.4 GaAs wafer (schematically) with two flats which indicate the wafer orientation: The orientation flat (typically 32 mm for a 100 mm wafer diameter) and the identification flat (typically 18 mm). Notice that the $[1\bar{1}0]$ and the $[\bar{1}\bar{1}0]$ directions are not equivalent for a zinc blende crystal.

    - tempering of boules
- mechanical processing
    - crystal trimming and diameter grinding
    - flat grinding (about flat see below and Fig. 3.4)
    - wire sawing/slicing to get raw wafers
    - edge rounding
    - lapping
    - etching/cleaning
- end preparation
    - mechano-chemical polishing
    - cleaning
    - final inspection, packaging, certification
    - delivery to customer

After thorough preparation, most wafers are sold to the customer as *epi-ready*, i.e., no additional preparation or cleaning step is necessary before loading into the epitaxial apparatus.

Such wafers (or substrates) are a prerequisite for epitaxy (see next chapters). They typically act just as

- seed crystals and
- mechanical carriers.

whereas in most cases, they do not participate as active material in the device performance. The total typical thickness of the wafers is 300–1000 μm, while the total thickness of the epitaxial layers is in most cases below 5–10 μm.

Typically for GaAs and InP wafers, the (001) surface is prepared, in many cases with a well-defined slight mis-orientation. The rotational orientation of the wafers is marked by **flats** (see Fig. 3.4). Later, devices (e.g., laser diodes) are arranged along specific crystal orientations with the help of these flats. Owing to the lower symmetry of the zinc blende lattice as compared to the diamond lattice, the perpendicular $[1\bar{1}0]$ and $[\bar{1}\bar{1}0]$ orientations are not equivalent, which is important for some devices, e.g., GaInP laser diodes grown on GaAs wafers. This can be traced back to the fact, that in the zinc blende lattice, the two directions of the [111]

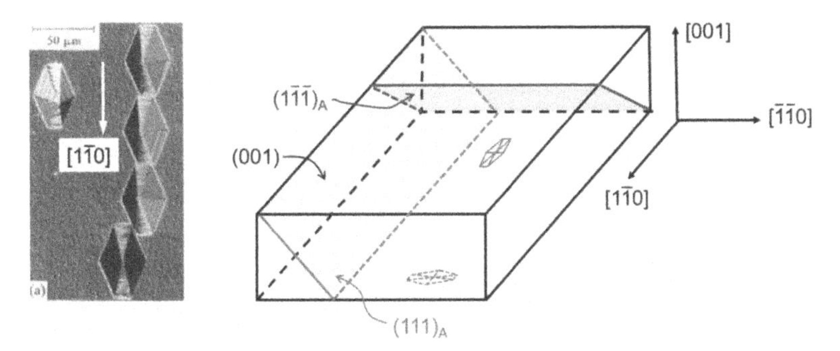

Figure 3.5 Orientation and shape of etch pits on a GaAs (001) surface. Left: Photograph of pits etched by molten KOH; right: Orientation and shape with respect to the crystal directions (Courtesy of J. L. Weyher, Polish Academy of Sciences). Notice that the ($(111)_A$) plane perpendicular to the [1$\bar{1}$0] direction faces upward, whereas the ($(111)_A$) plane perpendicular to the [111] direction faces downward.

orientation are not equivalent (c.f. Fig. 1.9): In one direction, a Ga atom is directly following an As atom, whereas in the opposite direction this is just reversed. Therefore, when preparing GaAs {111} surfaces, then we must distinguish between the $(111)_A$ surface (Ga-terminated) and the $(111)_B$ surface (As-terminated). By wet etching, typically the Ga-surface ($(111)_A$) gets exposed. In consequence, this means: If a hole is etched into the (001) surface of GaAs, then it typically has an elongated hexagon shape (Fig. 3.5).

Following are the important wafer properties:

- diameter of the wafer (typically measured in multiples of 1 inch)
- orientation
- defect density, typically the dislocation density
- surface quality, flatness, bow, warp
- electrical conductivity depending on the doping of the bulk crystal (n-type, p-type, semi-insulating)

Typical wafer diameters used in production are:

- GaAs: 4″ to 6″ (100–150 mm);
- InP: 2″ to 4″ (50–100 mm).

Problems

(1) Some questions about bulk crystal growth:

  (a) What is the main reason for growing GaAs bulk crystals?
  (b) What are the two main methods to grow GaAs bulk crystals? Explain how they work.
  (c) A major difference between these methods is the defect density in the grown crystals. Which kind of defects must be mainly considered? Which method produces typically a larger defect density? Why?
  (d) Which of these methods can be used to grow GaN bulk crystals? Explain.

(2) What is the difference between the $[\bar{1}10]$ and the $[110]$ direction on a GaAs wafer? Explain.

(3) Explain:

  (a) Why is the $\{110\}$ plane a good cleaving plane in GaAs and InP, whereas in Si, the best cleaving plane is the $\{111\}$ plane?
  (b) Why is it not possible to prepare a good $\{100\}$ plane by cleaving in a GaAs or InP crystal?

Hint: Fig. 1.9 might be helpful. Try to plot two-dimensional projections of the zinc blende cubic unit cell in which the proposed cleaving planes can be drawn as lines.

Chapter 4

Epitaxial Methods

As already mentioned briefly, most modern devices made of compound semiconductors contain epitaxially grown heterostructures. Therefore, **epitaxy** is a key technology for such purpose and hence gets particular focus in this book.

In Wikipedia, we find an explanation for this term: **Epitaxy** (Greek; "epi" meaning "on" and "taxis" meaning "in ordered manner") describes an ordered crystalline growth on a (single-) crystalline substrate. It involves the growth of crystals of one material on the crystal face of another (heteroepitaxy) or the same (homoepitaxy) material. Epitaxy forms a thin film whose material lattice structure and orientation or lattice symmetry are identical to that of the substrate on which it is deposited. Most importantly, if the substrate is a single crystal, then the thin film will also be a single crystal.

Nowadays, mainly two advanced methods are discussed in compound semiconductor science (metalorganic vapor phase epitaxy, see Section 4.3, and molecular beam epitaxy, see Section 4.4). However, here we will discuss a few more methods which are helpful for understanding some details and which still have their technical relevance as a production method for some cheap devices.

Compound Semiconductors: Physics, Technology, and Device Concepts
Ferdinand Scholz

ISBN 978-981-4774-07-9 (Hardcover), 978-1-315-22931-7 (eBook)
www.panstanford.com

4.1 Liquid Phase Epitaxy (LPE)

Historically, this is the pioneering technology. Technically, it just means crystalline growth from a (liquid) solution, which most of us know from the growth of, e.g., salt crystals out of a supersaturated salt-water solution. By changing the boundary conditions (typically by decreasing the temperature) the solution gets saturated and eventually over-saturated. You may find more details about this technology in [10, 11].

For compound semiconductor LPE, the solvent is a metal, in most cases one of the constituents of the compound, e.g., Ga or In. In this metal, the other components (e.g., As or P or dopants) are solved. The mixture is liquid at least at elevated temperature.

An important fact of epitaxy including LPE is, that crystal growth is now possible drastically below the melting point by supersaturation of the solution. This can be nicely understood by knowing the thermodynamic phase diagram of such a solution, see Fig. 4.1.

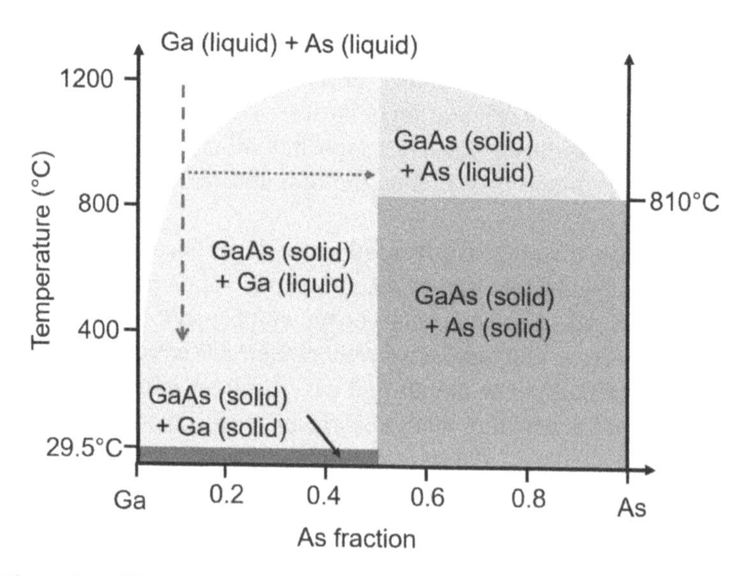

Figure 4.1 Phase diagram of GaAs. LPE is typically performed in a Ga-rich solution at temperatures around 700°C (i.e., on the left end of the figure). The solid always grows stoichiometrically (vertical line at 50%).

At high temperatures, the two constituents (in our example Ga and As) form a perfect mixture. When cooling down, some solidification may occur forming a solid which may have a different composition as the solution. We may consider as an example a solution with about 10% As solved in Ga (vertical arrow on the left hand side of Fig. 4.1). When the arrow representing the temperature decrease enters the "GaAs (solid) + Ga (liquid)" phase range (light brown area), solid GaAs (with an atomic ratio of 1:1) will form, as this is the only stable configuration for these elements. Hence the liquid solution and solid GaAs will co-exist. With further cooling, more GaAs grows. Depending on the amount of Ga-As solution, this may lead to a change in the composition of the solution, because relatively more As is required for GaAs out of a 10% solution.

As we see in this phase diagram, basically the same happens for all compositions on the left-hand side. The temperature, at which GaAs starts to grow, depends on the composition of the solution and decreases the closer we are at the Ga-rich corner. Therefore, the growth temperature of such GaAs is much lower than its melting point of 1240 °C (in the center of the diagram). Not directly visible in such a phase diagram is the growth velocity, which indeed decreases with temperature (see below).

For ternary systems, the relation is more complicated but can be calculated in many cases. One example for $Ga_xIn_{1-x}As$ is shown in Fig. 4.2, indicating that by the right choice of the composition of the solution and the temperature, a ternary material of given composition can be grown.

Some requirements to the solvent must be given:

- high solution potential for components
- chemical inertness
- low melting point
- high boiling point, low vapor pressure
- low viscosity
- weak wetting of crucible.

Notice: III-V compound semiconductors always crystallize (more or less) stoichiometrically, i.e., with a ratio of the group III atoms to the group V atoms of 1:1 (see Fig. 4.1).

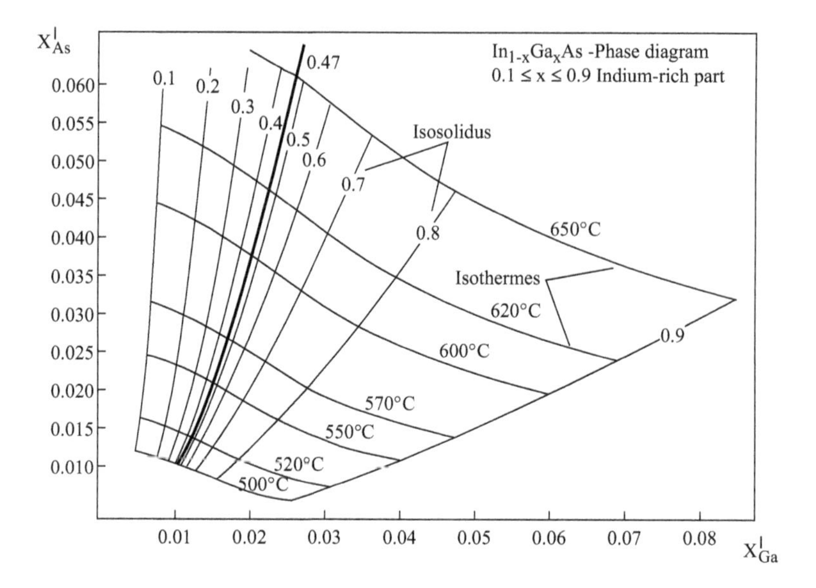

Figure 4.2 Phase diagram of the ternary system Ga-In-As. The solvent is In. The *isosolidus* lines indicate the composition of the ternary epitaxial layer from a solution of the given *liquidus* composition and the respective temperature (isothermal lines) after [12].

4.1.1 Control of Growth Rate and Layer Thickness

In order to start growth, basically the supersaturation of the solution is obtained by cooling.

There are different ways to do this:

(1) Ramp cooling or equilibrium cooling:
The process starts at $T = T_S$ (saturation point), then the temperature is reduced with a constant rate.
The layer thickness is then

$$d \sim \frac{\Delta T}{\Delta t} \cdot \frac{1}{C_S \cdot m} \cdot D^{1/2} \cdot t^{3/2},$$

with t = growth time, C_S = concentration of solved material (e.g., As in Ga solution), m = gradient of liquidus curve (K/mol), D = diffusion constant of solved material in solvent (e.g., As in Ga). This is because As has to be transported from the "bulk" solution to the growing surface.

Notice: Thickness does not grow linearly with time, the growth rate is proportional to $t^{1/2} \neq$ const.!

(2) Step cooling:
The solution is cooled to a temperature below T_S resulting in a supersaturation (without any crystallization yet), then it is brought in contact with the substrate which acts as a nucleation crystal. Crystalline growth starts thus reducing the supersaturation.
The thickness now is governed by

$$d \sim \Delta T \cdot \frac{1}{C_s \cdot m} \cdot D^{1/2} \cdot t^{1/2}$$

(3) Supercooling:
Start with step cooling, then continue with ramp cooling.

The formulas are only exactly true for semi-infinite solutions. They are good approximations as long as $t < l^2/D$ (with l = dimension of solution) and for

- diffusive transport, no convection;
- No nucleation in solution, i.e., not too large supersaturation; for GaAs, ΔT should be below 10 K, for AlGaAs typically below 2.5 K.

If the concentration of one constituent in the solution is very small, then the solution changes its composition substantially during growth, which makes the composition control more difficult.

Doping of the crystal can be done by mixing the dopant element into the solution (e.g., Si in Ga). Here, a small distribution coefficient $(\frac{\text{dopant in solid}}{\text{dopant in liquid}})$ is better keeping the concentration control feasible.

4.1.2 LPE Equipment

Typically, the growth in LPE is timed by mechanically moving the metallic solution onto the substrate.

This is often done by using

- a tilting crucible or
- a sliding boat technique.

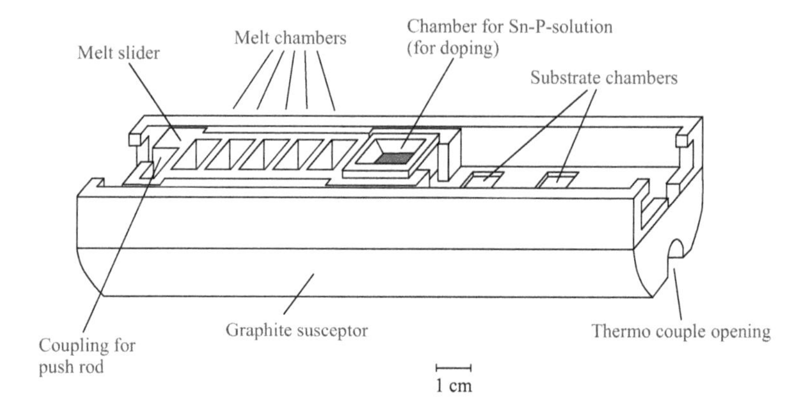

Figure 4.3 Sliding boat susceptor of a LPE system (after [13]).

As the latter is better suited for the growth of multi-heterostructures, it is explained in more detail here.

In the sliding boat technique, a multi-chamber sliding boat (Fig. 4.3) is positioned in a kind of guard rail susceptor, in which the substrate is fixed. The chambers can be filled with different growth solutions. These can be moved over the substrate one after the other thus enabling the growth of a multi-heterostructure. The layer thickness is controlled by the temperature profile and the time as discussed above. The whole susceptor is positioned in a quartz tube (Fig. 4.4) floated with high-purity hydrogen[a] in order to keep all environmental impurities including oxygen away from the growing semiconductor. Besides purity, the most critical parameter to be controlled is the temperature, measured by a thermocouple.

A typical process runs as follows:

(1) Mixing of the carefully out-weighted materials, filling of chambers
(2) Heating/annealing of solutions for purification
(3) Determination/cross-check of saturation temperature by visual inspection during slowly cooling down

[a]Hydrogen is used because it acts as an inert gas and moreover it can be purified easily by, e.g., a Pd diffusion purifier.

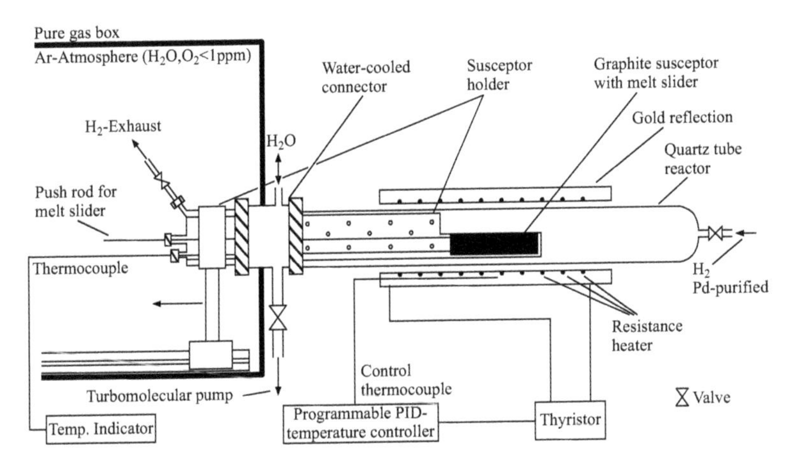

Figure 4.4 LPE system (after [13]). The resistance heater can be shifted to the right to allow fast cooling of the susceptor.

(4) Growth (step, ramp or super cooling). Typical data for GaAs: $T = 600°\text{C}$, $\Delta T = 3 - 6$ K, $t = 10$ min, resulting in a layer thickness of 2–5 μm.
(5) Cooling down to room temperature

Because of the non-linear relation between time and growth rate, it is very difficult to grow very thin layers by LPE. Moreover, the growth in such short times is not yet in equilibrium, therefore the composition of ternary layers may be not correct. With a high amount of care, thicknesses down to 10–20 nm could be achieved. However, this problem is the main reason that LPE is not applied for most modern devices which require the controlled growth of very thin layers (see Chapter 8).

Another basic problem is that earlier grown layers may be solved again by the solution of a later grown layer (*Melt-back*). This drastically limits the material combinations in hetero structures.

LPE is a very simple **equilibrium** growth method relying strongly on the thermodynamic phase diagram of the respective compound. Therefore, some multi-component compounds cannot be grown due to miscibility gaps which may depend on temperature (Fig. 4.5). Indeed, such immiscibility problems may also be important in other growth methods (see later).

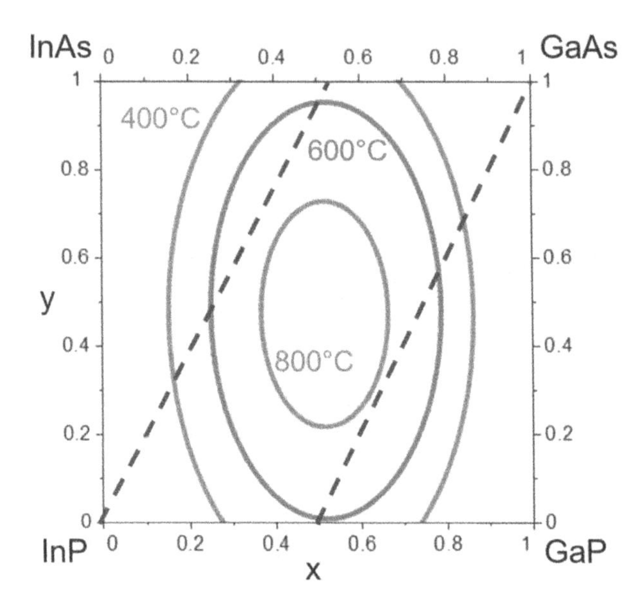

Figure 4.5 Miscibility gap in the quaternary system $Ga_xIn_{1-x}As_yP_{1-y}$. At the indicated temperatures, the material inside the ellipses cannot grow (after data from [14]). The blue broken lines mark those compositions which are lattice matched to GaAs or InP.

Due to these limitations, LPE is not well-suited for the growth of sophisticated device heterostructures, although it can produce highest quality materials.

However, it is still in use for the mass production of simple and cheap devices such as standard infrared LEDs. In this case, multiple full wafers are immersed into the liquid solutions.

4.2 Vapor Phase Epitaxy (VPE)

In order to overcome the basic problems of LPE, i.e.,

- limitations in layer sequence,
- control of growth rate, and
- phase diagram as boundary condition,

and to simplify the control of the growth process, the method of vapor phase epitaxy was invented. Its basic idea is: Control the supply of precursor material to the growth zone by controlling gas flows.

However, the elements which are needed (Ga, Al, In, As, P) are not gases, but in nearly all cases solids!

Therefore, chemical compounds are necessary which may be gaseous and which should release the respective element (and only this element) in the growth process. This works quite well for group V elements which form gaseous hydrides like AsH_3, PH_3, and NH_3.

Unfortunately. metals typically do not form gaseous compounds at all! Way out: A volatile compound can be formed at high temperatures in situ in the epitaxial reactor. This can be done by reacting the metal with HCl gas at temperatures above ca. 600°C.

For III-V compound semiconductors, two main methods have been developed:

- Hydride VPE (Tietjen method) using hydrides as group V precursor gases;
- Chloride VPE (Effer method) generating HCl in situ from other less critical chloride compounds.

Here, we concentrate on the Tietjen method with the much stronger technical relevance today mainly because of its use to grow thick GaN layers (see Chapter 11). Here, group V hydrides (as already mentioned) act as precursor gases, as they have suitable properties, e.g.,

- AsH_3: BP = –62°C, p@20°C = 1.5 MPa
- PH_3: BP = –88°C, p@20°C = 3.5 MPa
- NH_3: BP = –33°C, p@20°C = 857 kPa

They are available in compressed gas cylinders (either pure or mixed in hydrogen). A major problem of these gases is their extreme toxicity (see Section 4.3).

Now, the wafer on which the respective epitaxial layer should grow is positioned on a *susceptor* within a reaction chamber, again typically formed as quartz tube. HCl and the group V hydride gas are supplied to this tube by separate gas precursor lines. Hence, the control of group III and group V is independent from each other (see Fig. 4.6).

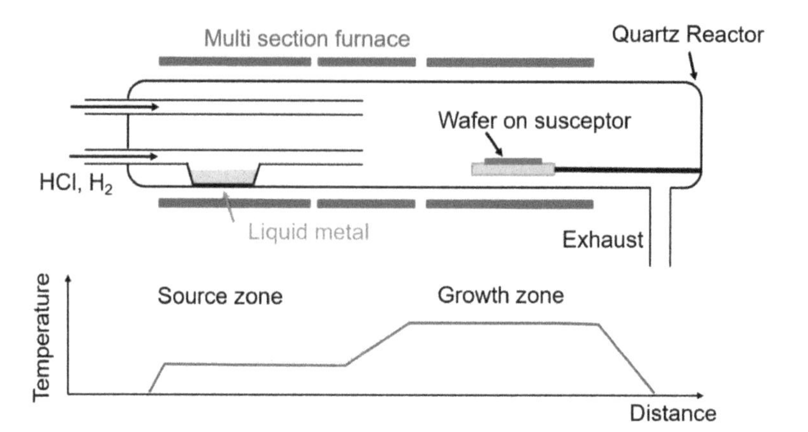

Figure 4.6 Top: Hydride VPE system, bottom: Temperature profile (schematically).

The reaction tube is heated from outside (hence this is a "hot wall reactor") to temperatures in the range of several 100°C. Near the gas inlet, the required elemental metal is stored in a quartz container. The incoming HCl gas reacts in situ in the reaction chamber with the hot metal forming $MeCl_{Gas}$ which further flows to the substrate thus transporting the metal component to the growth zone.

At the substrate surface (heated typically to 600–800°C), the growth reaction takes place:

$$6\ MeCl + X_4 \rightarrow 4\ MeX + 2\ MeCl_3$$
or
$$3\ MeCl + X_2 \rightarrow 2\ MeX + MeCl_3$$

X denotes the group V element (As, P), which forms X_4 or X_2 after the respective hydride XH_3 is thermally cracked.

Typically, a specific temperature profile is realized (Fig. 4.6 bottom) in such a reactor by a multi-zone heater. Between the metal source and the susceptor, a hot intermediate zone is necessary in order to crack the fairly stable hydride (see Fig. 4.6).

In general, VPE requires comparably simple and cheap equipment. Anyway, high-purity material can be achieved with high growth rates of several 100 μm/h. However, it is not very flexible in terms of the growth of heterostructures. Typically, only single binary layers can be grown, whereas the growth of ternary layers or respective heterostructures is not suitable with this method.

Hence, this method is used to fabricate simple, cheap LEDs, Gunn elements, detectors, field-effect transisters, i.e., simple devices where material purity is essential, or for the growth of thick layers (e.g., GaN, see Chapter 11).

4.3 Metalorganic Vapor Phase Epitaxy

Metalorganic vapor phase epitaxy (MOVPE) is nowadays the most relevant epitaxial method for III-V-compound semiconductor devices. Therefore, excellent textbooks exist describing many more details as we can do here, e.g., [15–17].

Besides the term "MOVPE" which is heavily used in Europe and which emphasizes its character to provide some real *epitaxial* growth, not just disordered deposition, other abbreviations are used synonymously:

- MOCVD = metalorganic chemical vapor deposition, mainly used in the United States
- OMVPE = organometallic vapor phase epitaxy (indeed chemically more correct)
- OMCVD = organometallic chemical vapor deposition

This method was invented in the late 1960s of last century by H. M. Manasevit (see, e.g., [18], but also the discussion in [15]) to make use of all advantages of VPE, but overcome its problems.

So let us summarize (as discussed in the last section):

On the one hand, VPE is potentially very flexible. If all source materials are available as gases, then the precursor supply is easily controllable:

- gas flow through tubes → geometrical distance not a limit
- switching with valves → fast and simple switching
- quantitative control of gas flow with "regulators," i.e., electronic mass flow controllers.

This enables the simple fabrication of

- different (binary) materials;
- multi-component layers;
- heterostructures;
- doped layers;
- abrupt or graded interfaces.

However, on the other hand, the following main problem must be overcome to make full use of these promising features:

There are no gaseous compounds available for group III elements (metals).

The solution is to use **metalorganic** (or organometallic) compounds, which, although being in many cases liquid, sometimes solid, have a fairly high vapor pressure.

The typical example is tri-methyl-gallium TMGa = $Ga(CH_3)_3$, which has a melting point M.P. = –15.8°C, a boiling point B.P. = 55.8°C and a vapor pressure p_{vap} = 8560 Pa at 0°C (see also Fig. 4.8).

In order to use such material as a gaseous precursor, it is stored in a so-called bubbler (Fig. 4.7) at a stabilized temperature which defines its vapor pressure p_{vap}. Through this bubbler, a carrier gas is flown which gets saturated with the material according to its vapor pressure. Assuming that all gaseous materials behave as ideal gases, then the transported amount of material can be derived from

$$n_{MO} = \frac{f_{H_2} \cdot p_{vap}(T) \cdot p_0}{p_{Bubbler} \cdot R \cdot T_0}, \qquad (4.1)$$

with n_{MO} = amount of transported material (in moles/min), f_{H_2} = flow of carrier gas (typically hydrogen) through bubbler (in m^3/min), p_0 = standard pressure (10^5 Pa), $p_{Bubbler}$ = pressure in bubbler, R = gas constant, and T_0 = standard temperature (15°C).

In many cases, the vapor pressure of a liquid can be described by

$$\log p(\text{Torr}) = B - A/T \qquad \text{or} \qquad \ln p(\text{Pa}) = B^* - A^*/T$$

and the constants A and B are specific material properties. Some data are shown in Fig. 4.8.

Notice: In the gas lines on the way to the epitaxial reactor, any recondensation due to oversaturation of the precursor material in the carrier gas must be avoided. Therefore, the bubbler temperature should be below room temperature (as the gas lines are at room

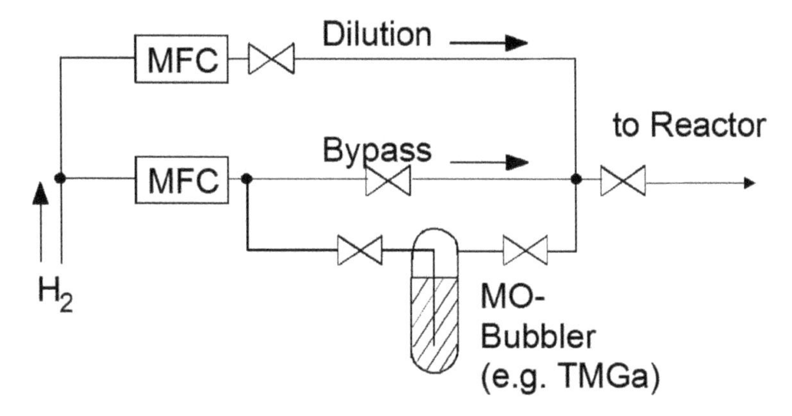

Figure 4.7 Liquid precursor channel. The amount of carrier gas flowing through the bubbler is controlled by an electronic MFC (mass flow controller). The carrier gas flow can be switched from a bypass line to the bubbler in order to keep the carrier gas flowing also in stand-by mode. Notice: The final supply to the reactor is controlled by a three-way valve (see Fig. 4.9).

temperature). Moreover, the precursor gas may be diluted by adding more carrier gas after the bubbler (see Fig. 4.7).

In order to be useful for the metalorganic vapor phase epitaxial process, the MO precursors should fulfill the following basic requirements:

- adequate vapor pressure (see below)
- preferably liquid, not solid in bubbler
- chemically stable during storage and in bubbler
- chemically stable in gas phase when mixed with other precursor gases
- thermal cracking at hot substrate to release the target element
- high purity in order not to contaminate the semiconductor
- release only the wanted element, all other should result in non-reactive gases going into the exhaust
- physical and chemical safety (in best case non-pyrophoric, non-explosive, non-toxic)
- acceptable price

In order to achieve an acceptable growth rate of 1–5 μm/h, a minimum vapor pressure of about 100 Pa is required. Typical carrier

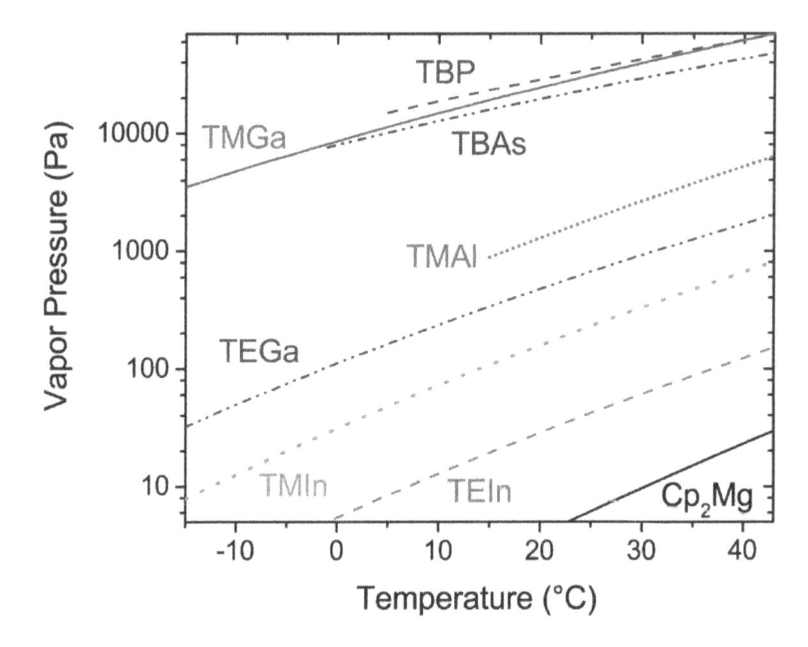

Figure 4.8 Vapor pressure data of some metalorganic precursors.

gas flow in normal bubblers is limited to about 200 sccm,[b] leading to some tens of micromoles per minute transported to the reactor. If the vapor pressure is too low, then a reduction of the bubbler pressure can help (see Eq. 4.1) to increase the molar flow.

To fulfill the requirements of vapor pressure and purity, the smallest organic molecules containing only hydrogen or carbon are most appropriate. Therefore, the metal methyls are best suited (like TMAl, TMGa, TMIn). For some elements, ethyls (e.g., TEGa, $Ga(C_2H_5)_3$) are acceptable (see Fig. 4.8).

Elements such as O, S, and Cl would lead to a contamination in the epitaxial layers and should therefore be avoided in the MO molecule.

Rule of thumb: The vapor pressure is related to the size of the molecule, i.e., the size of the alkyl group and/or the size of the metal atom.[c] Therefore, it decreases when going from the methyl

[b] sccm = standard cubic centimeters per minute.

[c] Notice the fairly low vapor pressure of TMAl. This molecule is a dimer in the gas phase, i.e., it evaporates not as $Al(CH_3)_3$ but as $(CH_3)_3Al - Al(CH_3)_3$.

to the ethyl compound. In parallel, the bond strength between the alkyl group and the metal decreases, thus the cracking efficiency increases. The latter fact may also result in less carbon incorporation in the epitaxial layer for ethyl precursors. The lower stability, however, may be also a critical fact which may lead to storage problems and to parasitic reactions in the gas phase (which is particularly true for TEIn).

Following is a short discussion of the metalorganic precursors of the most important elements:

- Gallium: TMGa is well suited in the light of the above-mentioned problems. Sometimes, its vapor pressure is regarded as too high making the control of small flows (required for, e.g., $Al_xGa_{1-x}As$ with large x) very difficult. Then, TEGa can be used instead.
- Aluminum: Al as well as TMAl is chemically very reactive. Therefore, oxygen contamination is a big problem. Also, C uptake may be a problem due to high bond strength of Al-C in TMAl. Alternative Al precursors have been discussed, but are typically not used today.
- Indium: The purity of In compounds (TMIn, TEIn) has been an issue making them extremely expensive (>50 Euro/g). The main problem with TMIn is its high melting point of 88°C. Hence, it has to be used as a solid, filled as a powder into the bubbler. Special bubbler constructions have solved most problems related to this feature. TEIn as possible alternative suffers from a low vapor pressure and low chemical stability. Other alternatives have been investigated. However, TMIn is still the best choice.

Similarly, doping metal elements can be transported to the reactor (see below).

With these metalorganic group III precursors and the group V hydrides, all components for the complete vapor phase epitaxial growth of III-V semiconductor layers are available. The precursor gases are transported to the hot substrate where they react (in a simplified picture) according to the following equation:

$$Ga(CH_3)_3 + AsH_3 \rightarrow GaAs + 3CH_4$$

Thus, the idea of a totally gas-flow controlled VPE system can be realized.

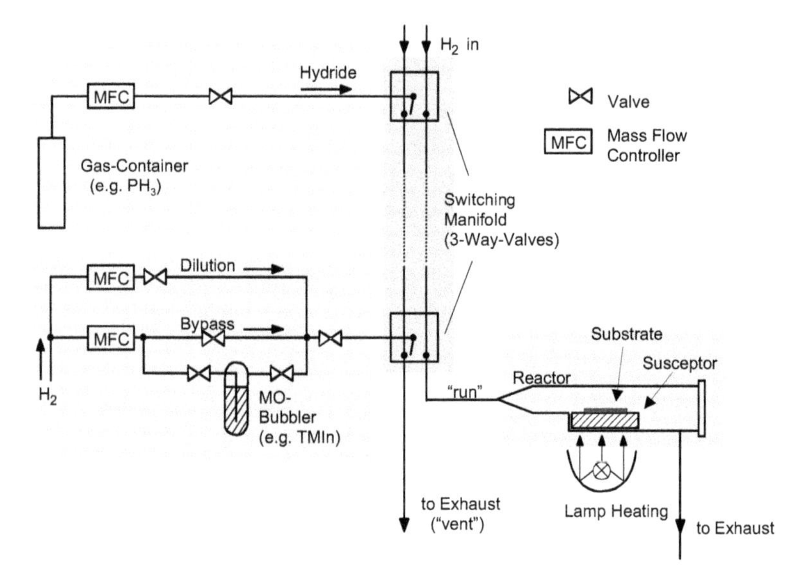

Figure 4.9 MOVPE system (schematically). More gas and MO channels can be added.

A complete MOVPE system contains the following main building blocks (Fig. 4.9):

- Source channel(s) for metalorganics providing a carrier gas flow loaded with the appropriate amount of precursor material (see also Fig. 4.7).
- Source channel(s) for gases providing the adequate hydride gas flow.
- Gas switching manifold controlling which gases enter at which time the reactor. So-called three-way (or five-way) valves enable the switching of a steady gas flow from an exhaust line into the reactor line to provide stable flow from the very first moment. This gas switching manifold is very important for the controlled growth of extremely thin layers with abrupt interfaces.
- Reactor where the epitaxial growth takes place (Fig. 4.10). In order to define the appropriate flow pattern, the reactor is fed, besides the precursor gases, with a larger amount of inert gas (typically hydrogen, sometimes nitrogen). More details see below.

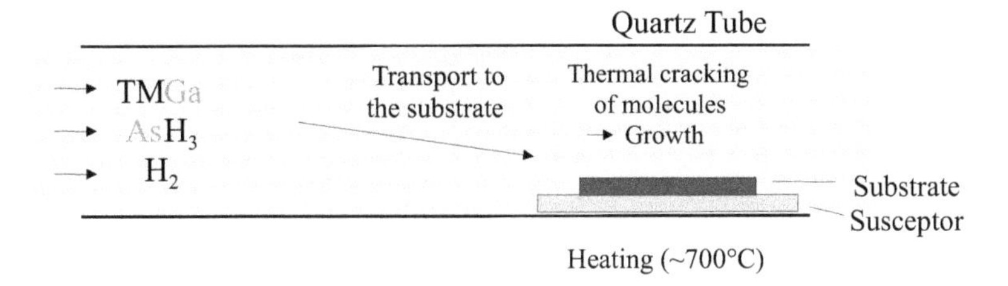

Figure 4.10 MOVPE reactor (schematically, growth of GaAs as example).

- Control unit (computer, electronics, etc.) providing control about the gas flows, the switching valves, the temperature, etc.
- Vacuum system (pump + controller) to stabilize the pressure during growth. See discussion below about epitaxial growth at reduced pressure.
- Exhaust system where the exhaust gases are treated to remove any toxic fractions before releasing it to the atmosphere.

The epitaxial growth (see Fig. 4.10) takes place on the substrate, i.e., an adequate wafer (c.f. Chapter 3) acting as a nucleation crystal. It is placed on a susceptor which can be heated from outside. Typically, the susceptor is made of graphite, coated with SiC to have a chemically inert surface. Heating is typically done either by RF induction or by lamp heating, i.e., only the susceptor is directly heated. Therefore, MOVPE is a **cold wall reactor** method (whereas LPE and in particular HVPE is a hot-wall system). Hence, the thermally induced chemical reactions of the precursor gases take place only near the substrate or on the substrate surface.

When measuring the growth rate r as a function of susceptor (or growth) temperature, 3 regions can be identified (Fig. 4.11):

(1) Low temperatures: The chemical reaction rates limit the growth rate r. As the reaction rates depend on temperature according to an Arrhenius law:

$$k = e^{-\frac{E_A}{kT}},$$

the growth rate also follows such an Arrhenius law, which gives a straight line in an Arrhenius plot ($\log(r)$ versus $1/T$).

(2) Medium temperature (roughly between 700 K and 1200 K depending on the materials): Growth rate does not depend

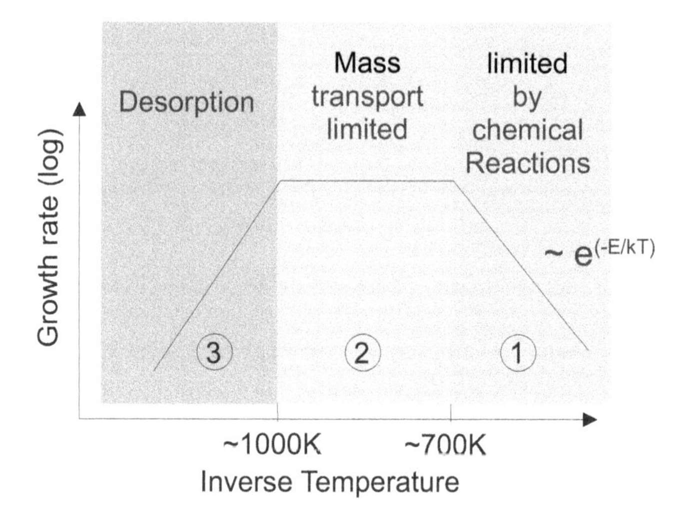

Figure 4.11 Log(growth rate) over inverse temperature (schematically).

on temperature, because now it is limited by the group III precursor supply. This is the normal temperature range for MOVPE. The primary layer properties (e.g., thickness, composition) depend nearly exclusively on the precursor flows. This region is called **Mass transport limited region**.

(3) High temperatures: Thermodynamic factors like desorption or dissociation of the crystal lead to a decrease of the growth rate with increasing temperature.

Indeed, the so-called **V/III ratio** is a very important growth parameter. Under nearly all circumstances, the group V precursor gases are supplied in excess (V/III $\gg$ 1). Reasons:

- Hydrides are fairly stable. Thus they may be cracked only partly even at high temperatures.
- Group V elements desorb easily from the hot layer surface, if no excess group V gas partial pressure is present.
- The adsorption Θ of the group V elements can be described by so-called *Langmuir isotherms*:

$$\Theta = \frac{kp}{1+kp},$$

with p = partial pressure, k = reaction constant. Even for large values of kp, only one monolayer is adsorbed ($\Theta \to 1$), then the deposition saturates.

For group III, the deposition would just continue resulting in a thick metallic layer or in metallic droplets, (if no group V element is available or if the V/III ratio is too small).

Consequently, in the mass transport limited region, the growth rate in MOVPE is (more or less) governed by the group III precursor supply.

How do the precursor molecules reach the substrate surface? The so-called **boundary layer model** helps to understand this process in a simple picture. Above the substrate, a **boundary layer** forms (see Fig. 4.12). Above the boundary layer, the precursor gas concentration in the total gas flow is defined by the partial pressure ratios according to the precursor flows. Any (group III) precursor molecule arriving at the substrate will be incorporated into the growing epitaxial layer. Thus the concentration is zero here. In the boundary layer, the molecules move to the substrate driven by diffusion as a result of the concentration gradient (see Fick's laws of diffusion). The diffusion constants depend only weakly on the precursor molecule types. Therefore, the amount of group III atoms arriving at the surface is directly related to their gas phase concentration above the boundary layer, i.e., to the precursor gas flow. This is why now the growth rates is mass transport limited (see Fig. 4.11).

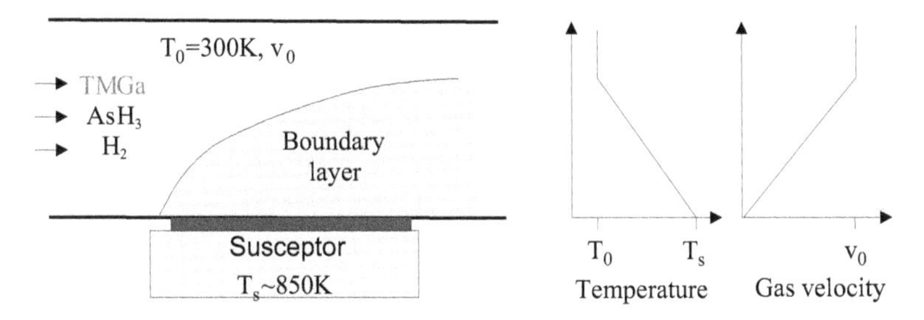

Figure 4.12 Boundary layer in a horizontal MOVPE reactor (schematically, left). Over the substrate, a temperature and gas velocity gradient forms (right).

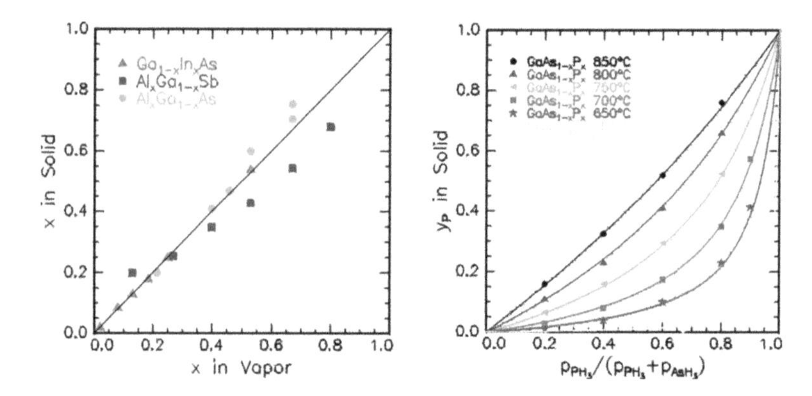

Figure 4.13 Composition in the solid as function of the gas phase composition (after Stringfellow [19], from the *Journal of Crystal Growth* with permission from Elsevier). Left: Group III mixing, right: Group V mixing.

This makes the control of the composition of ternary layers with two group III elements (III-III-V compound) quite easy: In many cases, the composition in the solid is exactly the same as the ratio of the partial pressures of the precursor gases in the reactor, not depending on temperature or other "secondary" epitaxial parameters (Fig. 4.13 left). Some exceptions (e.g., GaInN) will be discussed later (see Section 11.5).

For group V mixing, it is more complicated: Here, the composition in the solid depends, besides the group V gas flows, strongly on the temperature and other parameters (Fig. 4.13 right) as a consequence of the higher and temperature-dependent stability of the hydrides (see above), making the control of some V-V compositions very challenging. Example: To achieve larger values of y in $GaAs_{1-y}P_y$, an extremely high PH_3 flow is required due to the much larger thermal stability of PH_3 as compared to AsH_3.

Dopants can be supplied in the same manner as the main matrix elements, either as metalorganic liquids or as hydrides.

Typical dopants in III-V compound semiconductors are (see Chapter 5):

- n-type: Either hydrides of group VI elements as H_2Se, H_2S replacing group V elements, or group IV element (Si in GaAs, etc.) as SiH_4 replacing group III elements.

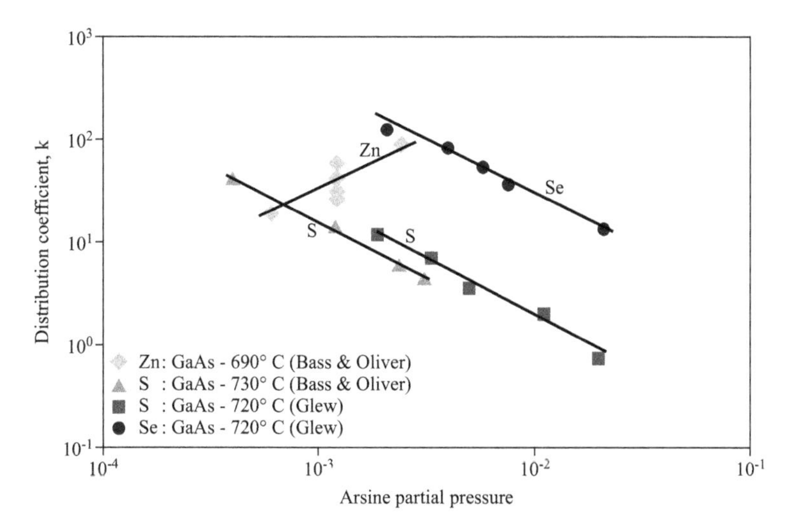

Figure 4.14 Distribution of different dopants as function of the V-III ratio represented by the arsine partial pressure (after Stringfellow [20], from the *Journal of Crystal Growth* with permission from Elsevier).

- p-type: Typically group II elements as metalorganics like di-methyl-Zn ($(CH_3)_2Zn$) or bis-cyclo-pentadienyl-Mg (Cp_2Mg) replacing group III elements, also group IV element (C in AlGaAs) replacing group V elements.
- Deep levels for semi-insulation: e.g., Fe in InP as $Fe(C_5H_5)_2$ (ferrocene).

Depending on which element is replaced, the doping efficiency may increase or decrease with changing V/III ratio (Fig. 4.14).

Big problem: Toxicity

As already briefly mentioned, the hydrides used as group V precursors are highly toxic with TLV levels in the range of 50–300 ppb.[d] Moreover, they are stored in high-pressure cylinders. Hence, the gas spreads over a large area in case of an accident.

[d] TLV = threshold limit value, the gas concentration which is accepted as still not dangerous for workers; ppb = parts per billion, 1 per 10^9.

This requires well-suited countermeasures to guarantee workers' safety in research and production:

- Construction:
    - All gas lines, etc., are made of high-quality pipes, valves, etc., typically made of stainless steel.
    - The whole MOVPE equipment needs to be checked regularly thoroughly concerning gas leaks.
    - The most critical lines are constructed as double wall pipes. By monitoring the shell volume of this co-axial structure, any leak of the inner or outer tube can be easily detected.
- Control: The gas concentration in the ambient air of the lab and in the double wall tube is continuously measured with respect to smallest traces of toxic hydrides.
- Scrubbing of exhaust gases in order to remove all toxic gases before releasing to the atmosphere:
    - Dry scrubbers: Adsorption and chemisorption by charcoal or other compounds;
    - Wet scrubbers: Chemical binding of the toxic gases in liquids by forming, e.g., group V salts;
    - Burning of exhaust gases in a hydrogen flame.

What could be done else? In the 1990s, strong research efforts have been done for the development of safer group V precursors, which may have

- lower toxicity and/or
- lower vapor pressure.

Such behavior is expected from "metalorganic" group V precursors. While compounds like TMAs ($(CH_3)_3As$) resulted in low-quality layers with strong carbon doping, best results are obtained with tert-butyl compounds (e.g., H_2–As–C–$(CH_3)_3$, TBAs or the same with P: TBP).

These compounds are liquid at room temperature and can be handled like the above discussed MO compounds, i.e., by using the bubbler technology (c.f. Fig. 4.7). They are still fairly toxic due to the remaining group V-H bonds. However, owing to their significantly lower vapor pressure, the risk of accidental release of the material is strongly reduced.

Another significant advantage of these compounds is their lower pyrolysis temperature and consequently their more efficient pyrolysis, allowing lower growth temperatures and/or V/III ratio.

Such materials, although they never could replace the hydrides in a larger fraction, are in use for some specific processes, in particular if low growth temperatures are required.

Reactor design aspects

An important component of an MOVPE machine is of course its reactor, where the epitaxial growth takes place. Hencc, its design is a key factor in the MOVPE growth process.

The following basic requirements are obvious:

- High growth rate, high precursor efficiency;
- Deposition on large areas (depending on application, e.g., solar cells, huge areas would be mandatory);
- However, good uniformity for all layer properties (e.g., thickness, composition, doping) over such large areas (single wafer or multiple wafers) are definitely required;
- The growth of abrupt hetero-interfaces and well-defined thin layers must be possible.

Over the recent decades, several quite different reactor geometries have been developed, which all fulfill these requirements to a large extend:

- Typical research reactor: Horizontal quartz tube with horizontal gas flow (Fig. 4.15). Problem: Such a reactor provides inherently a non-uniform growth rate due to the depletion of the precursors, i.e., incorporation of precursor material at the upstream end of the susceptor, decreasing the precursor concentration at the downstream end.

  Two countermeasures help to overcome this problem (and are also applied in the other reactor geometries, which are mentioned below):

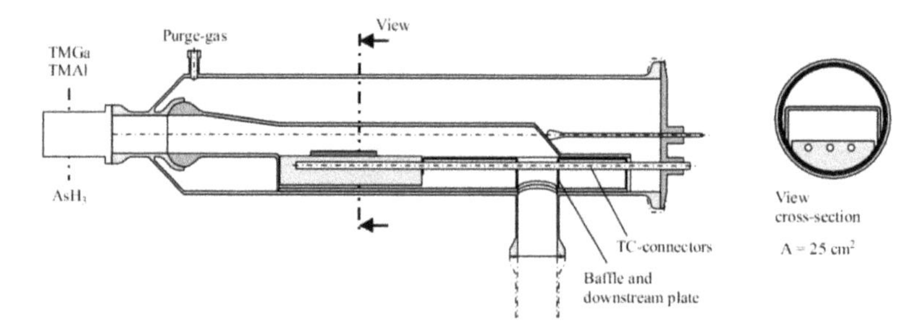

Figure 4.15 Horizontal MOVPE reactor with inner liner tube (see cross section right) providing rectangular cross section, which helps to improve the lateral layer homogeneity (reprinted from [21] with permission from Elsevier).

  - Reducing the gas pressure in the reactor (see extra paragraph below) and
  - By rotating the wafer during growth, the non-uniform growth rate gets averaged, leading to excellently homogeneous layers. Typical rotation speeds are in the range of one revolution per second.

- For larger areas and/or many wafers in industrial device production, planetary reactors for several wafers are used. These reactors are also classified as "horizontal" reactors with the gases entering the reaction chamber in the center and flowing radially (horizontally) over several wafers positioned around the center axis. Each wafer rotates by its own axis, additionally all wafers rotate around the central axis. (Fig. 4.16).

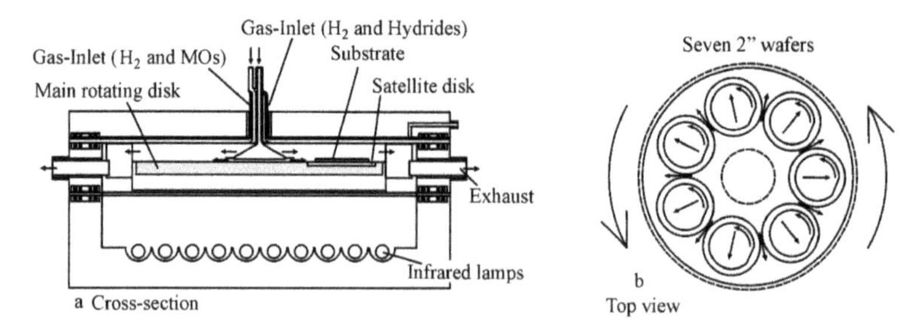

Figure 4.16 Planetary MOVPE reactor (reprinted from [22] with permission from Elsevier). Every satellite is rotated by gas foil rotation, additionally the main susceptor disk rotates.

Figure 4.17 Planetary MOVPE reactor opened for loading of the wafers (Aixtron AIX 2800G4 Planetary Reactor® 15 × 4 inch; with kind permission by AIXTRON SE).

Today, commercial systems for several wafers with large diameters are available, e.g., five wafers with a diameter of up to 200 mm or 60 wafers with 50 mm diameter (see, e.g., Fig. 4.17).

- Also reactors with vertical gas flow share a significant part of the MOVPE market. A very widespread design uses a "shower head" over the wafer(s) as gas inlet, which allows to optimize the local gas flow by different radial zones (Fig. 4.18). Hence, excellent layer uniformity can be obtained even for large area substrates with diameters of 8″ (20 cm).

In all cases, the reactor chamber is designed so that the gas flow remains laminar, because any turbulence would degrade the interface quality in hetero structures.

Reactor pressure

Typically, the gas pressure in an MOVPE reactor during growth is somewhere between 25 hPa and atmospheric pressure.

Figure 4.18 Showerhead MOVPE reactor opened for loading of the wafers (Aixtron AIX R6 Close Coupled Showerhead® Reactor 121 × 2 inch; with kind permission by AIXTRON SE).

Low pressure is a challenge for the pumping system, because it should be maintained at quite large gas flows (20 slm[e] and more). Moreover, corrosive gases must be pumped.

Why low pressure anyway?

- Historically, the main reason was the suppression of parasitic gas phase reactions between different precursor molecules. Some less stable MO precursors easily react with the group V hydrides already in the gas phase far away from the susceptor. Such parasitic reactions can be suppressed by increasing the mean distance between these molecules by low pressure. About 4 decades ago, this was the key for the successful MOVPE growth of InP.
- Moreover, lower pressure results in a higher gas velocity, which helps to get
    - abrupt interfaces in heterostructures;
    - a more homogeneous growth rate over larger areas;
    - laminar flow even for a not perfect reactor geometry.

---

[e]slm = standard liter per minute.

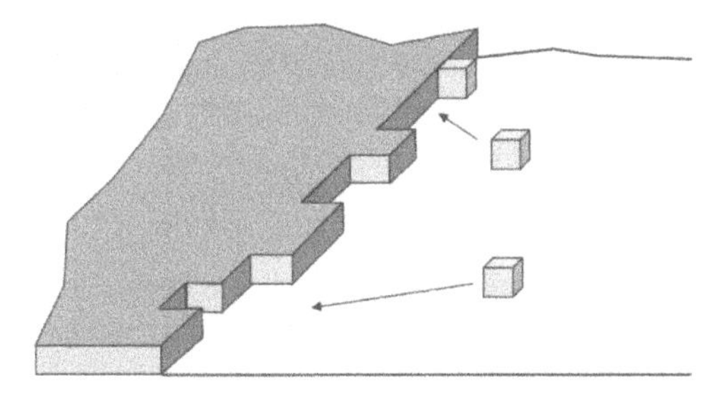

Figure 4.19 Atoms arriving on the growing surface typically move to an edge, because there their incorporation is energetically favored.

Growth of thin layers, etc.

In MOVPE (as in other epitaxial methods) the growth can be done in a parameter range where so-called "two-dimensional layer-by-layer growth" takes place. Atoms arriving at the surface have enough thermal energy to migrate on the surface until they find their correct site to be incorporated. This is typically at a surface step (Fig. 4.19). The surface is thus covered by moving the step over the total area. The next monolayer only starts when the first is finished (Fig. 4.20).

This growth mode is the pre-condition for the growth of atomically abrupt interfaces and hence the growth of nanometer thin layers.

Another pre-condition for abrupt interfaces is of course an abrupt switching of the precursor gases. This is technically realized with the above-mentioned three-way valves (see Fig. 4.9). Moreover, a laminar flow in the reactor tube must be established. Any turbulence would intermix the abruptly switched gas fronts.

To improve the interface abruptness and separate the material flows of different layers, growth interruptions may be introduced

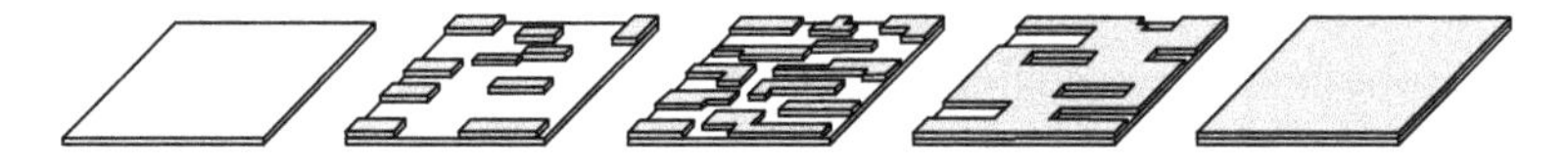

Figure 4.20 If the correct growth parameters are chosen (e.g., temperature, growth rate), atomic monolayers grow one after the other (after [23], reprinted with permission from Elsevier).

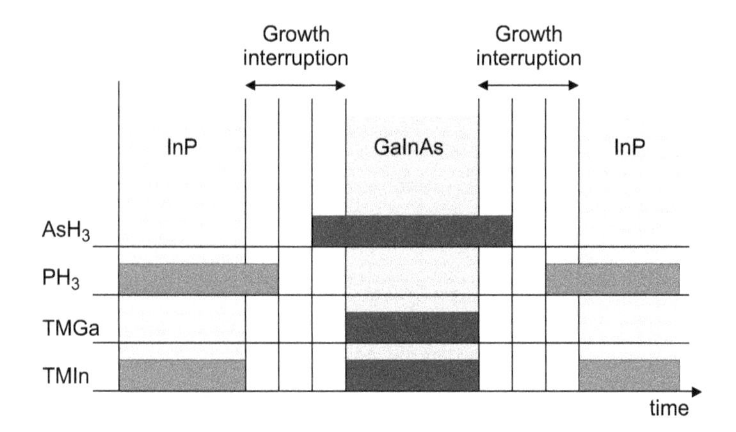

Figure 4.21 Growth interruptions for the growth of an InP-GaInAs double heterostructure. Depending on other growth parameters (e.g., reactor pressure), the interruption times have a length of some tenths to a few seconds.

(Fig. 4.21). For longer growth interruptions, it may be necessary to stabilize the surface by an ongoing group V precursor flow.

Nowadays, it is no problem to grow excellent layers with well-defined, uniform thickness in the range of 2 nm. The abruptness of thin layer interfaces can be directly seen in, e.g., photoluminescence experiments (c.f. Fig. 4.22).

In situ characterization

Due to the reactive gas phase, an in situ characterization is more challenging in MOVPE compared to high vacuum methods like molecular beam epitaxy (see Section 4.4).

However, excellent in situ characterization tools have been developed over the recent years, mainly based on optical methods:

- Simple reflection measurement: This enables to get information about surface roughness (intensity of reflected light) and layer thickness (due to multi-layer interference).
- Reflection anisotropy spectroscopy (RAS) or reflection difference spectroscopy (RDS, see Fig. 4.23): By analyzing the polarization of the reflected light r along the two lateral

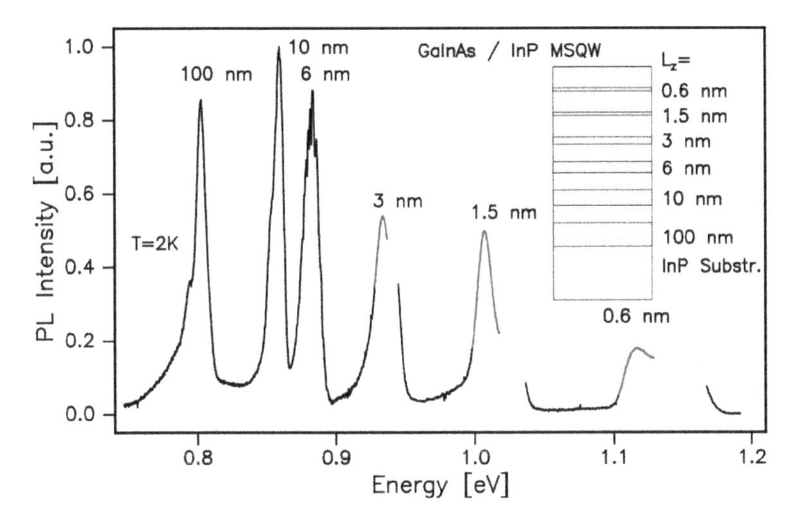

Figure 4.22 Low-temperature photoluminescence (PL) signal of a GaInAs-InP multi-heterostructure containing several quantum wells (QWs) with different thickness as described in the insert. As discussed in Chapter 8, the effective band gap of a QW and hence its PL line position depends on its thickness. Therefore, each QW gives rise for one PL peak at different energy. The signals of the thinnest QWs show some line splitting indicating that these QWs have very abrupt interfaces with excellent homogeneity over the radius of an exciton fluctuating by exactly one crystalline monolayer.

directions x and y in a wide spectral range (e.g., 250–850 nm),

$$\frac{\Delta r}{r} = \frac{r_x - r_y}{r_x + r_y},$$

detailed information on the surface reconstruction can be obtained, which may depend, besides temperature, on the composition or doping of the semiconductor layer. Typically, the signal is quite small: $\frac{\Delta r}{r} \simeq 10^{-3}$. These methods are still mainly used in a research stage, but on the move to be applied in running systems.

- Pyrometric true temperature evaluation: Particularly for the growth of GaN and respective heterostructures, very high growth temperatures above 1000°C are required. However, owing to the quite complex construction of rotating susceptors, it is very difficult to measure the real surface temperature of the susceptor or even the wafer. Here, a pyrometric measurement

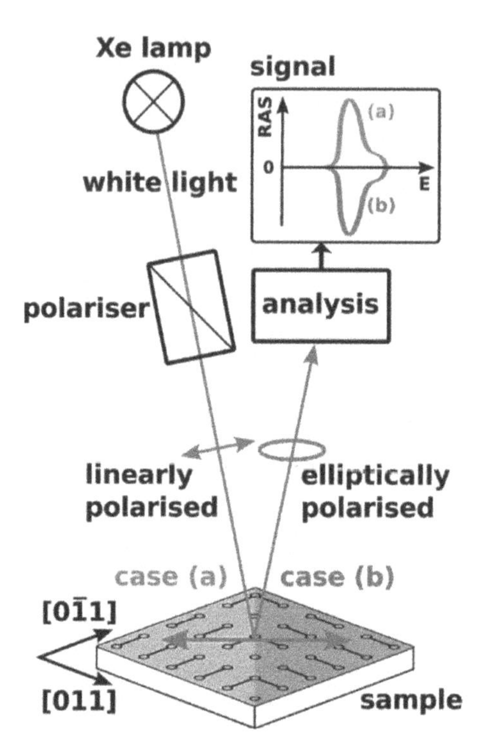

Figure 4.23 RAS setup for MOVPE (from [24], http://pubs.acs.org/doi/full/10.1021/jp502955m, with permission). Without the polarizer/analyzer filters, it may be used as a simple reflection measurement setup.

helps which is applied from top, giving excellent feedback to the operator about the real wafer temperature.

- Curvature measurement: In hetero-epitaxial systems like GaN on sapphire (see Chapter 11), the different mechanical and thermal properties of the layers and substrates result in strain (see Chapter 9) and consequently curvature of these structures. Hence, the curvature is a quantitative indication of strain. This can be measured in situ by analyzing the distance of two laser spots reflected on the growing layer surface.

Depending on the needs, integrated in situ characterization tools are available containing several or all these methods.

4.4 Molecular Beam Epitaxy

For compound semiconductors, the method of **molecular beam epitaxy** (MBE) is the second sophisticated method (besides MOVPE) for the growth of complex heterostructures for research and device applications. It was invented in the 1960s at Bell Telephone Labs. Again, we can only present a short description here. The interested reader can find more details in [25–28].

The basic idea of MBE is to perform "gas phase epitaxy" by using the needed elements and nothing else.

As the required elements for our compound semiconductors are not gaseous, but (nearly all) solids, they need to be vaporized, which requires high temperatures (some 100°C).

In consequence, a transport in gas lines is not possible (or the gas lines must be hotter than the vaporization temperature). Therefore, the materials are vaporized quite close to the substrate on which the epitaxial growth should take place, and the vaporized material is transported as molecular beam in an ultrahigh vacuum (UHV) environment. As a direct consequence, there is no problem with particle–particle interaction.

Hence the basic construction of an MBE system can be regarded as being quite simple (see Fig. 4.24): Just use an ultrahigh vacuum chamber. On one side, introduce molecular beams, and on the other side, place the substrate where the epitaxial growth should take place.

The key components of an MBE system (see Fig. 4.25) are

- evaporation sources (effusion cells);
- shutters which determine when a molecular beam is allowed to work;
- susceptor with substrate;
- substrate heating facility: Typical growth temperatures are somewhat (100–200 K) lower than in MOVPE;
- UHV chamber with pumps: typical growth pressures are in the range of 10^{-8} Pa.

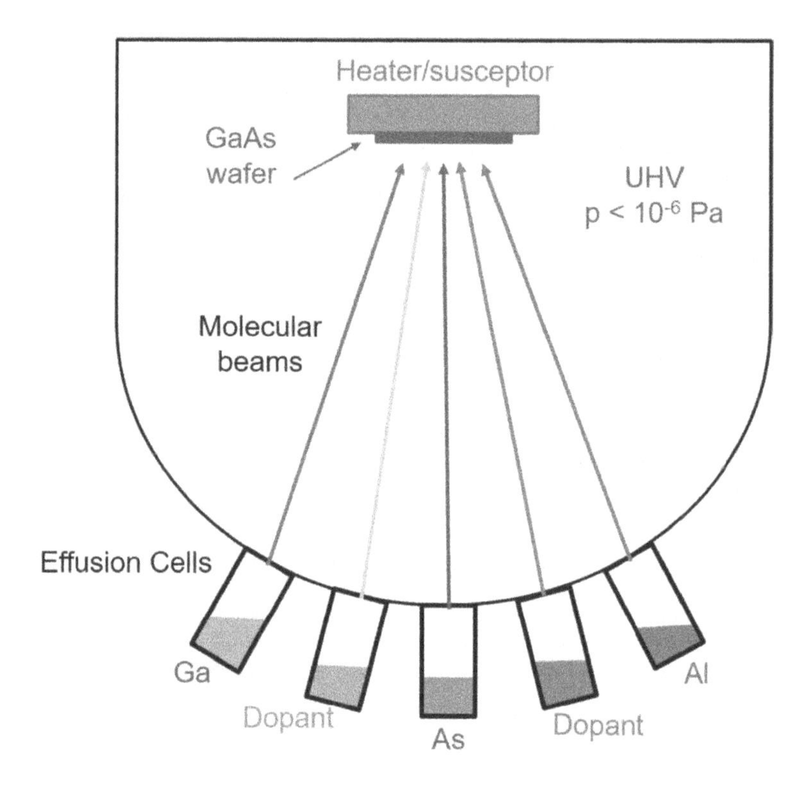

Figure 4.24 Basic principle of a MBE system: Every needed element is heated for evaporation. The respective molecular beams hit the substrate. Heating of the substrate is needed so that the atoms can find their correct site on the surface.

Following are some important facts about MBE:

- Quality of MBE layers is as good as or even better than that of layers grown by other methods.
- Surfaces and interfaces can be grown with excellent flatness (1 atomic layer or less). Exact thickness control is possible. The growth rate is in the range of 0.1 to 1 μm/h, i.e., 1/10 of a monolayer per second and less.
  The growth rate is in most cases (similar to that in MOVPE) mainly a function of the group III precursor supply. It does not depend on the substrate temperature.

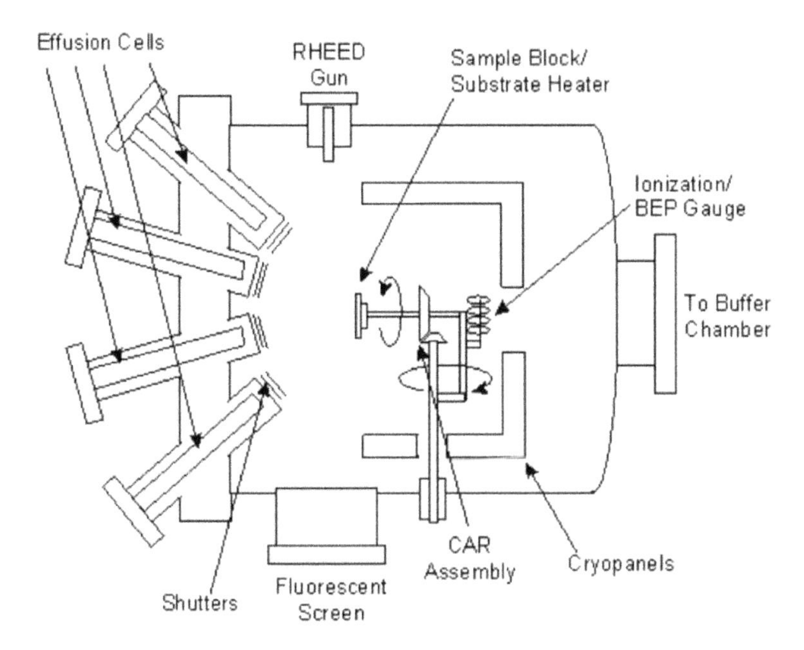

Figure 4.25 MBE system (schematically) (reprinted from [29], with the permission of AIP Publishing).

- An advantage of the UHV conditions is, that various in situ characterization methods can be applied easily, e.g.,
  - RHEED (reflection high-energy electron diffraction);
  - Auger spectroscopy;
  - ESCA (electron spectroscopy for chemical analysis).

  Even in situ processing is possible. The UHV chamber may be connected to other UHV chambers. Then even more methods are available such as UHV-AFM,[f] ion implantation, etc., without any contamination of the surface.

4.4.1 Equipment Details

UHV chamber: In order to obtain high-quality layers, strong requirements to leak tightness must be fulfilled in MBE. The

[f]Atomic force microscopy.

UHV conditions finally determine the purity of the grown layers. For acceptable layer quality, pressures down to 10^{-8} Pa must be achieved in comparably large volumes. Therefore, large stainless steel chambers with superfinished flanges are required. As pumping down from atmospheric pressure is a big problem, typically several chambers are connected serving as gate chambers for the loading of the substrate, etc. Hence, the ultra high vacuum in the growth chamber does not need to be broken completely when exchanging the substrate, etc. It is also possible to perform cleaning or pre-heating processes in the pre-chambers. The wafers can be transferred between these chambers without any contact to the environment, i.e., without the risk of surface contamination. To this end, sophisticated UHV manipulators are necessary along with big, but leak-tight transfer gates.

Pumps: Due to the required UHV conditions, pumps play a key role for the performance of MBE.

Depending on the pressure range, different pumps have to be used. The most important pump mechanisms are briefly discussed here:

- Rotary pumps: Mechanical pump for pressure range 10^5–10^{-1} Pa. For MBE, it should be an oil-free pump to minimize contamination by back-diffusion of oil.
- Roots pumps: Higher rotation speed, larger pump volumes than rotary pumps. Roots pumps have a typical compression ratio of about 100. Therefore, if combined with a fore-pump (rotary vane pump), they may reach end pressures of 10^{-2} Pa. Both mechanical pumps are mainly used as fore or booster pumps.
- Diffusion pumps: These pumps work without moving parts making them quite robust. Working principle (from Wikipedia): A diffusing vapor is generated by boiling a fluid (typically silicone oil). The vapor is directed through a multi-stage jet assembly. While diffusing, it transports residual gas molecules in the pump throat down into the bottom of the pump and out the exhaust. The outside of the diffusion pump is cooled using either air flow or a water line. As the vapor jet impacts the outer cooled shell of the diffusion pump the gas entrained in the jet flow coalesces, carrying the entrained pumped gases into the base of the pump where the gas pressure is increased

and pumped by the secondary mechanical or rough pump from the diffusion pump outlet. Diffusion pumps start to work at pressures below about 1 Pa, end pressures may reach 10^{-5} Pa and less depending mainly on the properties (vapor pressure) of the used oil. Main disadvantage: Back-diffusion of the oil leading to contamination of the pumped volume. This may be suppressed by a cold trap working with liquid nitrogen. As diffusion pumps are driven by convection, they have a low pumping efficiency. Such pumps are seldom in use at MBE systems.

- Turbomolecular pumps: From Wikipedia: These pumps work on the principle that gas molecules can be given momentum in a desired direction by repeated collision with a moving solid surface. In a turbo pump, a rapidly spinning turbine rotor 'hits' gas molecules from the inlet of the pump towards the exhaust in order to create or maintain a vacuum. Most turbomolecular pumps employ multiple stages consisting of rotor/stator pairs mounted in series. The mechanical energy and momentum of the extremely fast rotating rotor blades is transferred to the gas molecules. With this newly acquired momentum, the gas molecules enter into the gas transfer holes in the stator. This leads them to the next stage where they again collide with the rotor surface, and this process is continued, finally leading them outwards through the exhaust. As the momentum transfer depends on the mass of the molecules, a turbo pump works more efficiently with heavy molecules, less efficiently for He or H_2.

  The main advantage of turbomolecular pumps is that they have high pump efficiency and can reach excellent end pressures down to 10^{-8} Pa. The turbomolecular pump principle requires a small distance between the static and rotating blades smaller than the mean free path of the gas molecules. Therefore, these pumps typically need mechanical fore pumps going down to 1–10 Pa. As they work completely mechanically, they are well suited for high purity applications. In MBE, they are used for evaporation of the fore chambers, etc., but their end pressure is not low enough for the main chamber. However, extremely high rotation speeds of 20,000 to 90,000 rpm are requested making the pump very sensitive to any vibration or other fault (e.g., back-pressure due to power failure). Any such failure typically ends in a total destruction of the pump.

- Sublimation pumps: With sublimation and getter pumps, extremely low end pressures can be obtained. A typical example is the titanium (Ti) sublimation pump, where a Ti layer is evaporated by a hot filament and deposited on a cooled area. Other molecules are chemisorbed by reacting with the chemically reactive Ti deposited on the cold surface. After a while, new reactive Ti must be deposited. Active gases such as CO_x, O_2, and H_2O are easily trapped with high pumping rates. Intermediate gases such as N_2 or H_2 are trapped fairly well at low temperature (77 K). Noble gases are not pumped at all!
These pumps can be switched on only intermittently, as they need some regeneration after a while.
Anyway: They are very simple, clean and silent (no moving parts). Their pressure covers the range between 10 Pa and 10^{-9} Pa. Hence, they need (mechanical) fore pumps.
- Ion getter pump: It works similar to the Ti sublimation pump, where the evaporated atoms additionally get ionized. Therefore, they can be directed efficiently by electrical and magnetic fields. By creating and accelerating ions, also the gas molecules to be pumped out get ionized and thus are trapped more efficiently on the cathode. We can distinguish four mechanisms:

    - adsorption
    - gettering by freshly sputtered cathode material, i.e., chemical binding of the gas molecules;
    - surface burial under freshly sputtered cathode material
    - ion burial after ionization in the discharge region

  Here, also noble gases can be pumped with, however, still low efficiency. The pressure ranges from 10^{-2} to 10^{-8} Pa.
- Cryo pumps: Their main working principle is the condensation of material on a cooled surface. Somewhat more specifically, their pump mechanisms are:

    - Cryocondensation: All gases which may get liquid or solid at low temperatures (typically liquid nitrogen) get trapped at a respectively cooled surface.

- Cryosorption in cooled charcoal or molecular sieves, i.e., material with extremely large surface area. This is the only pumping mechanism for gases such as He, Ne, and H_2.
- Cryotrapping by dynamic trapping one gas on and in a porous frozen condensate of another adsorbed material.

The pressure range of cryo pumps is from 10^{-4} to 10^{-9} Pa. This pumping scheme is extremely clean, simple, and not sensitive to high pressure failures. However, a regeneration is necessary after short use. Therefore, typically two pumps are used in parallel.

A good overview about the working principles of these pumps can be found in [30].

As already mentioned, an extremely good vacuum below 10^{-8} Pa at room temperature is required for MBE, which is obtained by combining several pumps and particularly using sublimation and ion getter pumps. During the epitaxial process, cryo pumps are also in use. Turbo pumps are typically used for the evacuation of the fore chambers, e.g., after loading the substrate. Turbo and diffusion pumps in combination with cold traps are in use in MBE-related methods such as chemical beam epitaxy (CBE, see later), where the vacuum must be maintained in the presence of larger gas flows, but the end vacuum is substantially lower (10^{-4} Pa).

When growing GaAs, the walls of the vacuum chamber get covered with more and more As which would increase the back-pressure to 10^{-7} Pa or even higher. By cooling the chamber walls or at least the most critical parts (cryo shroud) with liquid nitrogen, the required better pressure can be achieved. Hence, liquid nitrogen consumption is one of the economic key issues of MBE.

After opening the chamber, which is necessary only in the case of maintenance, precursor refill, or some problem, it needs long time to pump down to the best pressure due to water adsorption from the environmental atmosphere. Then typically all chamber walls need to be heated to 200–250°C for at least 10–100 h, before the growth experiments can be continued. Thus after any such maintenance, a big, time-consuming effort is required to put the system back into operation (days or even weeks).

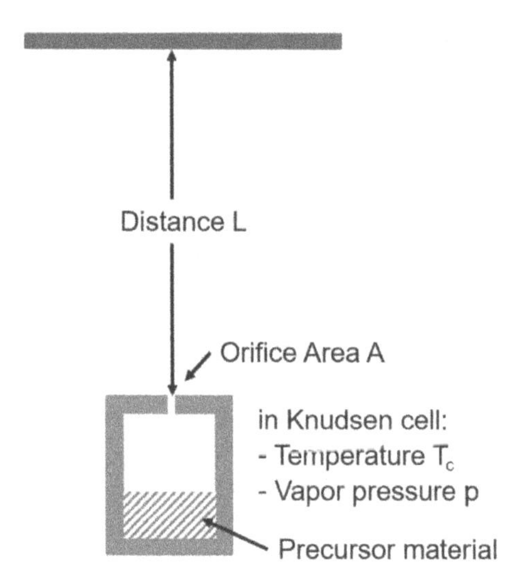

Figure 4.26 Knudsen cell (schematically).

Effusion cells: The precursors to be vaporized are put into effusion cells connected to the main MBE chamber.

The evaporation rate out of an ideal effusion cell should be stable over time, only depending on temperature, but not on the filling state of the cell. This can in principle be fulfilled by a **Knudsen cell** (Fig. 4.26): This cell has only a small orifice, so that the pressure in the cell is equal to the equilibrium vapor pressure of the material, whereas outside UHV conditions are present. The vapor pressure and hence the evaporation rate are only defined by the temperature. The number of atoms Γ leaving the Knudsen cell can be obtained from

$$\Gamma = \frac{p \cdot A \cdot N}{(2\pi \cdot M \cdot R \cdot T_c)^{1/2}},$$

with A = area of orifice, p = equilibrium vapor pressure, $N = 6 \times 10^{23}$, M = atomic weight of precursor, R = gas constant, and T_c = cell temperature.

This results in a deposition rate

$$J = \frac{\Gamma}{\pi \cdot L^2} \quad \text{(atoms per cm}^2 \text{ and sec)}$$

for a distance L between Knudsen cell and substrate.

Example: For a Ga effusion cell with $A = 0.6\,\text{cm}^2$ and $T_\text{c} = 1000°\text{C}$ (which results in a Ga vapor pressure of $p_\text{Ga} = 0.1\,\text{Pa}$) we obtain about $J = 3 \times 10^{15}$ atoms/(cm^2·s) which is equivalent to a growth rate of GaAs of about 1 μm/h.

Today, the cells typically have larger orifice areas A to increase the evaporation rate or decrease the cell temperature. Such cells are no longer real Knudsen cells, but still provide a fairly constant evaporation rate independent of the filling state.

By controlling the cell temperature and thus the evaporation rate, the growth rate and the composition of ternaries is controlled.

Typical cell temperatures are between 250 and 1400°C. They should be controlled to an accuracy of ±1°C to keep the ternary composition stable.

Typical cell materials must be inert and stable up to very high temperatures, as fulfilled by molybdenum, tantalum, pyrolytic BN (pBN) or graphite.

To keep the UHV conditions, an excellent thermal insulation between cells and MBE chamber (which needs to be cold, see above) must be realized.

The growth is controlled by mechanical **shutters** opening or blocking the molecular beam out of the respective cell. A major problem is that the thermal balance between the hot effusion cell and the cold MBE chamber depends sensitively on the shutter position. To minimize the thermal feedback of the shutter, it may be arranged in some angle. Moreover, although being close to the hot effusion cell, the shutter should be cold so that nothing can desorb from its surface.

By this shutter control, the evaporation is controlled more or less digitally: an effusion cell can be either completely "open" or completely "close." Therefore, graded compositions, where a smooth change of the precursor flow would be required, cannot be grown so easily in MBE. Obviously, MOVPE is much better suited for such gradings, as modern electronic mass flow controllers can be (more or less) instantly set to any new flow value. One chance in MBE is the ramping of the cell temperature. However, only small gradients around 40 K/min maximum can be obtained. Another way to grow graded layers is digital grading: By quickly switching

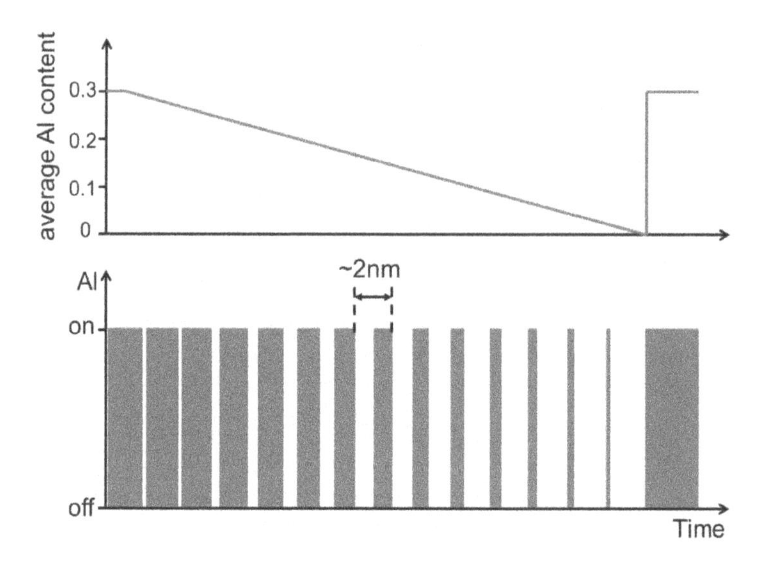

Figure 4.27 Growth of a graded ternary AlGaAs layer by fast switching of the Al shutter (lower diagram). The Ga and the As fluxes remain open all the time.

the shutters with variable on-off ratio (duty cycle), very short period superlattices can be grown, the average composition of which (depending on the duty cycle) can be regarded as ternary composition (Fig. 4.27).

Substrate holder: As in other epitaxial methods, also in MBE, a heating of the substrate is required. The main purpose of a hot substrate surface is to provide enough thermal energy to the adsorbed atoms to migrate on the wafer surface to find the most appropriate incorporation site (see. Fig. 4.19). Therefore, the mounting of the substrate on the susceptor is critical.

Different to other epitaxial methods, where the wafer is typically just laid on the hot susceptor and therefore is heated by the mechanical contact, this is not possible in MBE, because the molecular beams should come from below, i.e., the wafer must be positioned up-side down or at least vertically posted, because the effusion cells must be mounted with their orifice pointing to the top. Otherwise, the liquid precursors would flow out of the effusion cells. Therefore, a direct heat flow contact between wafer and susceptor is typically not given. Owing to the UHV conditions, any small gap between these two items acts as a very good thermal insulation.

In former days, the substrate was glued onto a Mo holder with In. For larger wafers, this is more and more difficult, because an inhomogeneous wetting results in an inhomogeneous substrate temperature.

Today, the wafer is typically mounted by some clamps or so without any In glue. Now, the heat is coupled to the wafer only by infrared radiation. Therefore, it is difficult to measure the substrate temperature by a thermocouple or some other direct sensor. The evaluated temperature may depend at least from the details of mounting and the substrate material.

One way out is a pyrometric measurement by evaluation of the black body radiation of the substrate surface:

$$S(\nu) = \frac{2\nu^2}{c_0^2} \cdot \frac{h\nu}{e^{\frac{h\nu}{kT}} - 1} \quad \text{Planck's law of radiation,}$$

with ν = frequency of IR radiation and c_0 = speed of light.

However, in particular for GaAs epitaxy, the temperatures are fairly low (around 500°C), hence the infrared signal of appropriate wavelength is very weak. Moreover, as the wafer (e.g., GaAs) is transparent for IR radiation, it may be difficult to obtain the signal of the substrate surface.

A more complicated, but more accurate method is to measure the absorption edge, i.e., band gap of the substrate (Fig. 4.28), which is a function of the temperature (Eq. 1.29), as discussed in Section 1.8.

Moreover, the substrate temperature can be calibrated:

- From time to time by measuring the melting point of InSb (527°C) or other compounds;
- By observing the desorption of the oxide on the GaAs wafer by RHEED (can be done in the beginning of every run).

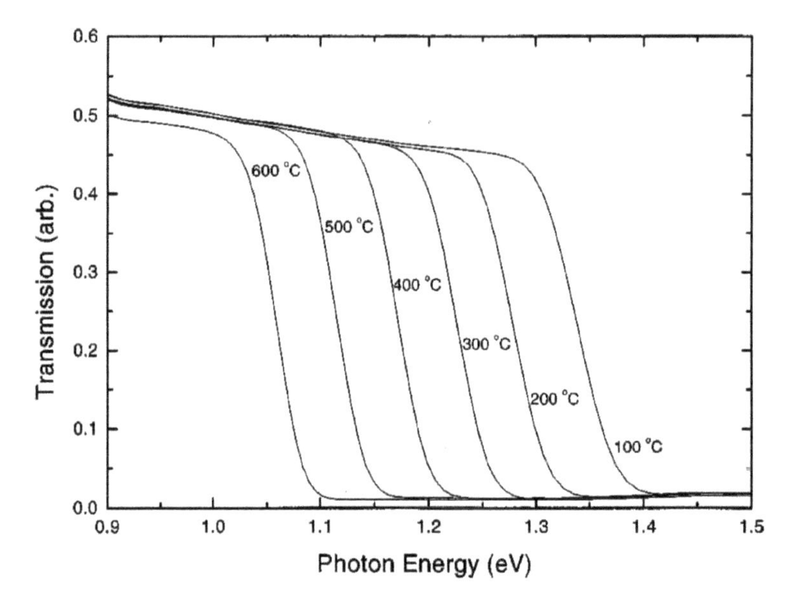

Figure 4.28 Absorption of GaAs at different temperatures (reprinted with permission from [31]. Copyright *J. Vacuum Sci. Technol. B* 1999, American Vacuum Society). The absorption edge is directly related to the band gap.

Then, the thermocouple signal obtained from the susceptor body can be used as an indication for the wafer temperature.

Like in MOVPE, the wafer may be rotated in order to obtain good homogeneity.

The intensity of the molecular beams can be evaluated by an ionization gauge which can be mechanically moved into the substrate position (see Fig. 4.25) to determine the **beam equivalent pressure** (BEP) as a measure for the growth rate (after adequate calibration).

In situ characterization: As already mentioned: Owing to the UHV conditions, in situ characterization is quite easy in MBE. The most important method (because in use at nearly every MBE machine) is **Reflected high-energy electron diffraction (RHEED)**: A focused electron beam (1–5 keV) hits the substrate (or growing epilayer) surface under a small angle (1–2°), where it is elastically diffracted. This is displayed on a fluorescence screen on the other

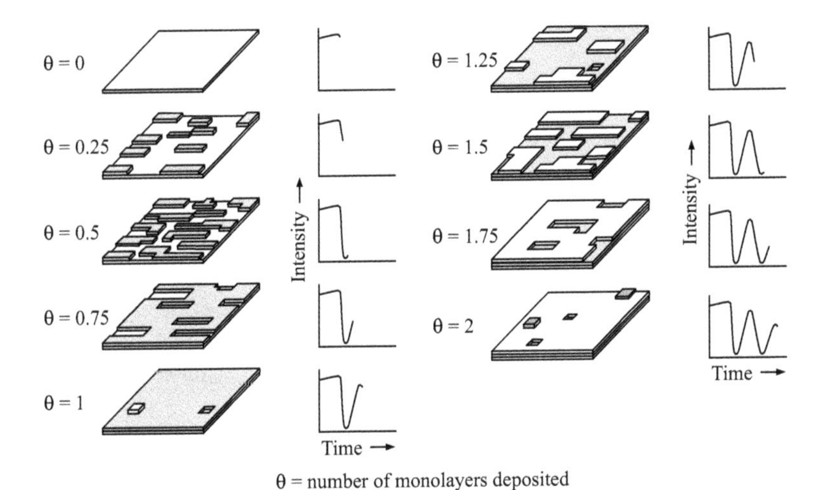

Figure 4.29 Growth of an epilayer: Under reasonable growth conditions, the growth proceeds monolayer by monolayer (left columns, c.f. Fig. 4.20). Completed, flat surfaces provide strong RHEED intensity, whereas incomplete atomic layers provide lower intensity (right columns) (reprinted from [23] with permission from Elsevier).

side of the MBE chamber (see Fig. 4.25). The diffraction pattern reflects the reciprocal lattice of the surface. Due to the small angle, only one or two atomic layers are probed—surface reconstruction can be obtained.

By this signal, the following details are detectable:

- Roughness of the surface: Only flat surfaces give higher intensity (Fig. 4.29).
- Surface reconstruction depending on the material properties and the growth environment (temperature).
- Surface properties such as oxide on GaAs and its desorption at higher temperatures. This information is useful to determine the start of the epi process (see above, temperature control).
- From the RHEED oscillations, the growth rate and layer thickness can be deduced in situ (Fig. 4.29). By counting the

monolayer oscillations, the layer thickness can be determined very accurately.

Indeed, many other in situ methods can be applied. One other method of major importance is **Auger electron spectroscopy**: Electrons hitting the surface may provide enough energy to kick out an electron in a deeper atomic orbit. Then another electron will relax to this orbit. It may transfer its energy to a third electron which then leaves the atom and the solid. The extra energy transferred to this electron is specific for every atom. By analyzing the kinetic energy of the released electron, a chemical analysis of the surface can be done.

Process-related features

- **Substrate preparation**: Typically, substrates (in particular GaAs) are provided "epi-ready" (c.f. Chapter 3). Thus no external preparation is required. However, some thin contamination layer from the environmental atmosphere may deposit on the surface, moreover, a native oxide may grow on the wafer. Therefore, before epitaxial growth is started, the substrate is heated in a preparation chamber connected to the MBE system at about 400°C in order to remove such contaminants. Then it is transferred in UHV to the growth chamber where it is heated to higher temperatures in situ to remove the native oxide. This is monitored by RHEED (see above).
- **Source preparation**: After filling the effusion cells with new source material, they are typically heated for a while to remove any contaminations (outgassing process). This may need long time.

Growth mechanism

For high-quality layers, typically a two-dimensional growth mode (step flow mode, c.f. Figs. 4.19 and 4.20) is established. Like in VPE, the group III precursor supply determines the growth rate.[g]

---

[g]Here, we discuss mainly "typical" materials and processes, e.g., growth of GaAs.

- The Ga sticking coefficient on the surface is about unity.
- As arrives as As_2 or As_4. It only dissociates at Ga atoms, otherwise it desorbs after short time. Thus, the As sticking coefficient is typically below unity: As has to be offered in excess. The V/III-ratio is larger than unity.

Doping elements are typically supplied as solid sources such as the matrix elements.

Doping atoms are first adsorbed at the surface to be either incorporated after a while or desorbed again. This results in a doping gradient even when the doping molecular beam is switched abruptly, depending on the sticking coefficient

$$s = \frac{k_\mathrm{i}}{k_\mathrm{i} + k_\mathrm{d}},$$

with k_i, k_d: incorporation and desorption coefficients. Best case: $k_\mathrm{i} \gg k_\mathrm{d}$.

This is fairly true for silicon (Si), the major **donor** for GaAs, where it is incorporated on a Ga site. Hence, abrupt doping profiles can be obtained. Additionally, the diffusion of Si in the GaAs crystal is very weak.

At higher temperatures (above about 600°C), surface segregation of Si on the GaAs surface takes place.

Other n-type doping candidates such as Ge, Sn, Te, S, and Se are less well suited. Some have too high vapor pressure leading to a quite strong desorption.

Best suited for **p-type doping** of GaAs is beryllium (Be), as its vapor pressure is well controllable, so that a large doping range between $p \sim 10^{15}$ and 10^{20} cm^{-3} can be easily adjusted. Additionally, it shows only weak diffusion and segregation and a fairly low activation energy of $E_\mathrm{A} \sim 19$ meV. However, it is highly toxic. This is why it is used only in MBE (inside a closed reaction chamber), but not in MOVPE.

Carbon, C, is a very good p-type dopant for AlGaAs. Due to its low vapor pressure, its use as solid source is very difficult. Typically, it is used as a gaseous source (CBr_4).

The handling of other p-type dopants such as Cd, Zn, and Mg is more difficult because of their volatility and their diffusivity in GaAs.

MBE-related basic problems

- Control of vapor pressure of some elements, e.g., phosphorus: High vapor pressure, high background pressure due to deposition on chamber walls; P is present in different modifications, which makes it very difficult to control its evaporation rate; Solution:
  - Grow only P-free structures (GaAs-AlGaAs, GaInAs-AlInAs);
  - GaP may be used as P precursor;
  - Supply the group V elements as hydrides like in MOVPE. The respective method is called **gas source MBE** (see below);
  - Specific two-step crackers, which can deal with the modification problems of P.
- In MBE, like in MOVPE, it is difficult to control the composition of ternary III-V-V compounds;
- Graded compositions as discussed above;
- UHV requirements making any maintenance very difficult (see above).

Many of these problems can in principle be solved by supplying the precursors as gases, which was a hot topic in research for a while about 25 years ago. The following sub-methods have been defined:

- **Metalorganic MBE (MO-MBE)**: Group III precursors supplied as MO gas channel like in MOVPE;
- **Gas source MBE (GS-MBE)**: Group V precursors supplied as gas channels (hydrides).
- **Chemical Beam Epitaxy (CBE)**: Both types combined (Fig. 4.30). This method seemed to be very promising some years ago, as it might combine the advantages of MOVPE (simple and exact flow control, simple exchange of precursors, most elements available) and MBE (molecular beams, monolayer

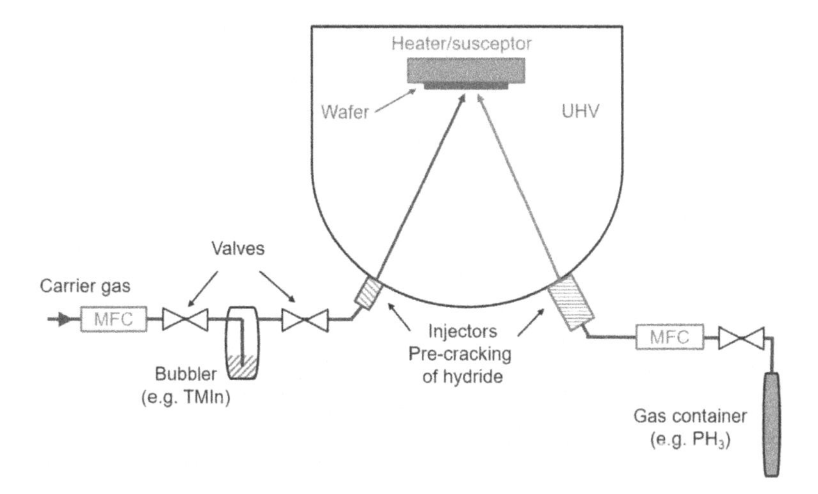

Figure 4.30 CBE system (schematically). Only two precursor channels are plotted, many more may be connected.

abruptness, no parasitic gas phase reactions). However, it did not really succeed.[h]

The only sub-method which turned out to survive is gas source MBE. By this method, the MBE growth of GaInAsP could be managed, a material class which is very important for long wavelength optoelectronic data communication. The hydrides AsH_3 and PH_3 are typically cracked in a high-temperature cracker cell before entering the reactor, so that only As_2 and P_2 travel in the UHV chamber.

The metalorganics, needed for MO-MBE and CBE, on the other hand, are cracked only on the hot substrate surface. This increases the (unintentional) carbon doping level, because other than in MOVPE, the formation of CH_4 is hindered because no reactive hydrogen is available.

Find more about these methods in [32, 33].

[h]Some people even mentioned that this method might combine the disadvantages of both methods.

4.5 Comparison of Epitaxial Methods

Let us briefly summarize at the end of the chapter the pros and cons of the different epitaxial methods in the following table (partly after [15]).

Method	Strengths	Weaknesses
LPE	Simple, high purity	Scale economics, inflexible
HVPE	Well developed, large scale, large growth rates, excellent thick GaN layers	No Al alloys, Sb alloys difficult, complex process/reactor, control difficult hazardous precursors
MOVPE	Most flexible, abrupt interfaces, high purity, complex layer sequence possible, simple reactor, robust process, uniform large scale, high growth rates, selective growth, in situ monitoring	Expensive reactants, most parameters to control accurately, hazardous precursors
MBE	Simple process, uniform, abrupt interfaces, excellent for GaAs/AlGaAs in situ monitoring	As/P difficult, Sb alloys difficult, N materials difficult, "Oval" defects, low throughput, expensive (capital)
GS-MBE	MBE advantages applied to GaInAsP (telecom), GaN	Low throughput, expensive (capital)
CBE	Uniform, abrupt interfaces, direct control of fluxes, selective growth	Low throughput, no large-scale reactors, expensive (capital), expensive reactants, hazardous precursors, N materials difficult

Problems

(1) (a) How can the growth temperature of a crystal be reduced substantially below its melting temperature?
    (b) What do you expect for the growth rate if the temperature is reduced?
(2) Explain the term "melt-back," which may occur in LPE.
(3) (a) Explain how hydride vapor phase works.
    (b) What is the main advantage of MOVPE compared with HVPE?
    (c) What is a major problem in conventional HVPE and MOVPE?
(4) (a) What is the typical temperature for the MOVPE of GaAs? What is the typical growth rate?
    (b) What is the typical temperature for the MBE of GaAs? What is the typical growth rate now?
    (c) Why do you need to heat the substrate for MOVPE and MBE?
(5) (a) What is typically used as a carrier gas in the MOVPE process of GaAs and related materials? Why?
    (b) Explain: What does "mass transport limited regime" mean?
    (c) How large is typically the V-III ratio in MOVPE? Why?
(6) (a) An MOVPE grower likes to grow GaAs. He sets the growth parameters as follows:

- Ga bubbler: Temperature $T_{\mathrm{Ga}} = 0°\mathrm{C}$, bubbler pressure $p_{\mathrm{Ga}} = 10^5$ Pa, carrier gas flow $j_{\mathrm{Ga}} = 7$ sccm;
- AsH_3: Flow $j_{\mathrm{As}} = 80$ sccm.

After 90 min growth, he obtains a 2 μm-thick layer of GaAs on a wafer with a diameter of 50 mm.
Calculate

- the V-III ratio during this run;
- the growth efficiency for the group III source.

(b) By additionally opening the In flow to the reactor, this scientist now likes to grow GaInAs lattice matched to InP. Which carrier gas flow through the TMIn bubbler is necessary to achieve such lattice matching composition? The TMIn bubbler is kept at $T_{\text{In}} = 17°\text{C}$ and $p_{\text{bubbler}} = 4 \times 10^4$ Pa.

(7) (a) Explain: Why is it more difficult to control the As-P ratio when you like to grow $\text{GaAs}_y\text{P}_{1-y}$ by MOVPE compared with the Ga-In ratio for $\text{Ga}_x\text{In}_{1-x}\text{As}$?

(b) Discuss some arguments why to use low pressure in MOVPE.

(8) (a) Explain: How can you grow multi-heterostructures in MBE?

(b) Explain briefly: How does an ion getter pump work?

(c) How can you grow a layer with graded composition by MOVPE, and by MBE?

(9) Can you measure the growth rate in situ in MOVPE and/or in MBE? Explain.

(10) (a) Which material class is difficult to be grown by MBE? Why?

(b) How can you overcome this problem? Explain.

(11) (a) Why is it more difficult in MBE compared with MOVPE to measure the wafer temperature accurately?

(b) Discuss some possibilities to measure the wafer temperature in the MBE process.

Chapter 5

Electrical Properties of Semiconductors/Electrical Characterization

5.1 Carrier Concentration and Doping

A significant feature of semiconductors is the fact that their conductivity can be drastically varied by different externals, particularly by doping, i.e., by mixing impurities of low concentration to the pure semiconductor material.

Remember (from Chapter 1.10, Eq. 1.36) that the temperature-dependent carrier concentration is related to the position of the Fermi level according to

$$n(T) = N_{\mathrm{C}} \cdot e^{\frac{E_{\mathrm{F}} - E_{\mathrm{C}}}{kT}}. \tag{5.1}$$

This means by changing the Fermi level E_{F}, we can manipulate the carrier concentration n. Vice versa, the inverse is also true:

$$E_{\mathrm{F}} = E_{\mathrm{c}} + kT \ln \frac{n}{N_{\mathrm{C}}} \tag{5.2}$$

How can we manipulate the carrier density n or the Fermi level E_{F}?

Compound Semiconductors: Physics, Technology, and Device Concepts
Ferdinand Scholz

ISBN 978-981-4774-07-9 (Hardcover), 978-1-315-22931-7 (eBook)
www.panstanford.com

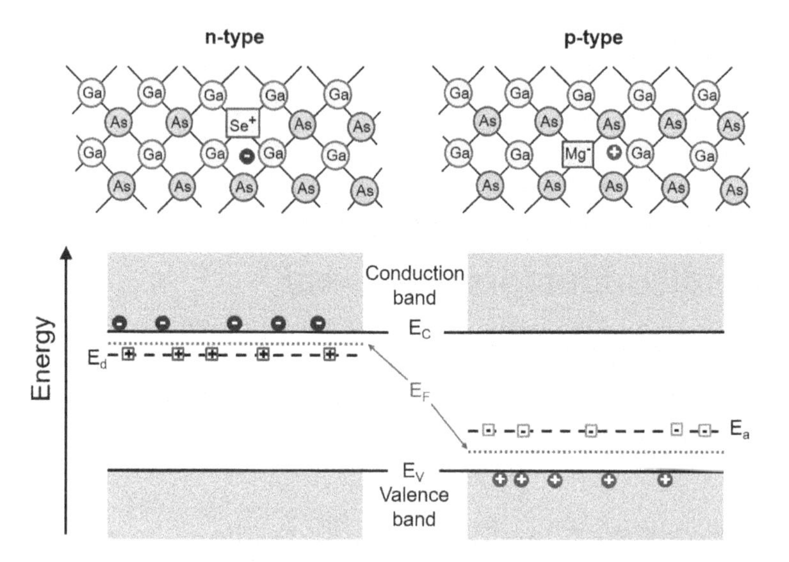

Figure 5.1 Doping of a compound semiconductor (schematically). Left: n-type doping with a group VI element (e.g., Se) replacing As in GaAs; right: p-type doping with a group II element (e.g., Mg) replacing Ga. The donors release their extra electron into the conduction band remaining as positively charged ions, whereas the acceptors catch an electron from the valence band turning into negatively charged ions thus leaving a hole in the valence band. Notice the position of the Fermi level close to the conduction and valence band, respectively, as it is the case for reasonable doping concentrations.

As briefly mentioned, this can be done by doping, i.e., by introducing impurities which provide extra carriers (Fig. 5.1). In III-V semiconductors, these may be elements of the group VI, e.g., S, Se, and Te. These elements have six instead of five valence electrons. They replace a group V atom in the crystalline structure; hence, they need to provide only five valence electrons for the chemical binding to their neighbors. The sixth electron contributes to an increased carrier concentration in the conduction band (see more details below). In some materials, Si and Sn as group IV elements replace the group III atom and thus act also as donor. This is in particular true for Si in GaAs and GaN.

As this extra electron is not needed for the chemical binding, it gets fairly freely movable in the whole crystal as an extra electron in the conduction band. However, some activation or ionization energy is necessary for this process, as the remaining donor atom gets a positively charged ion hence attracting electrons. The binding energy, i.e., the ionization energy can be calculated in the same way as the binding energy of an electron to a proton (forming a hydrogen atom), as also discussed in Section 6.2. From this model, we easily obtain the donor ionization energy E_D (see Eq. 5.3):

$$E_n = \frac{m^* e^4}{8h^2(\varepsilon_0 \varepsilon_r)^2 n^2} \quad \text{with } n = \text{quantum number} \tag{5.3}$$

For donors, this hydrogen model is indeed a very good description. In the hydrogen atom, the ground state energy is 13.6 eV. For our semiconductors, as the radius of the bound carrier is fairly large (at least about 10 nm), the crystal properties are averaged and taken into account in the dielectric constant ε_r. Moreover, we have to take into account the effective mass of the carrier (see Section 1.5). Therefore, the donor ionization energy is about $E_D = 6\,\text{meV}$ (for GaAs) fairly irrespective of the particular donor. This means the donor introduces a new energy level E_d slightly below the conduction band edge ($\Delta E = E_c - E_d = E_D$, see Fig. 5.1).

The carrier concentration can also be increased by doping with elements which have one valence electron missing for the chemical bonds, e.g., elements of the second column of the periodic table. If they replace a group III element, then they just take an electron from the valence band for the chemical bond, leaving an extra hole in the valence band. Typical acceptors are Zn, Mg, Cd, Be. Again, some group IV element may replace a group V element and hence acts as acceptor. Best example: C in (Al)GaAs and GaInAs.

As mentioned above, extra carriers means a change of the Fermi level position. At moderate doping concentrations, the Fermi level is just half-way between the donor level and the conduction band edge.[a] This means at $T = 0$, all extra electrons are still bound to their donors. For higher temperatures, the extra electrons can easily be excited thermally into the conduction band, which is governed by the same statistical rules as discussed for intrinsic semiconductors (Section 1.10). Now, the carrier concentration versus temperature reads (using the Boltzmann approximation, which is valid for small

---

[a] Most arguments are inversely true for holes with respect to the valence band.

Table 5.1 Ionization energies of some acceptors in different compound semiconductors (after different sources; some values are still under debate)

Binding energy (meV)	GaAs	InP	GaP	GaN
C	27	41.3	54	230
Mg	28.8	48	59.9	187-250
Be	10	14		250
Zn	30.7	46.6	69.7	340
Cd	34.4	57	102.2	550
Ca				(270?)

E_D and moderate temperatures):

$$n = \frac{N_\mathrm{D}}{2} e^{\frac{E_\mathrm{D}-E_\mathrm{F}}{kT}} = \frac{N_\mathrm{D}}{2} e^{-\frac{E_\mathrm{D}}{2kT}}, \tag{5.4}$$

where N_D is the impurity atom concentration.

As the donor ionization energy in most semiconductors is quite small (around 5 meV, see Eq. 5.3), donors are more or less completely ionized at room temperature. Then the carrier concentration is limited by the donor concentration (see Fig. 5.2). This means in some temperature range, the carrier concentration does not heavily change with temperature, which is quite helpful for the design of electronic devices, etc.

Due to the larger effective mass of the holes and the more complex nature of the valence band, the properties of the respective acceptors are more dominant; therefore, their ionization energy E_A is fairly specific to the different acceptors and has values in a range of 30 meV to several 100 meV (see Table 5.1). This even enables their identification by optical spectroscopy (see Chapter 6). As many values are substantially larger than the thermal energy kT at room temperature ($\approx$ 25 meV), the activation ratio of holes may be quite low. This is particularly an issue for GaN and related materials (see Section 11.4).

Please remember that the mass action law (Eq. 1.42) is still valid. This means if the electron concentration is significantly increased by doping, then the hole concentration decreases accordingly.

On the other hand, if both donors and acceptors are present in the semiconductor material, then compensation takes place: The

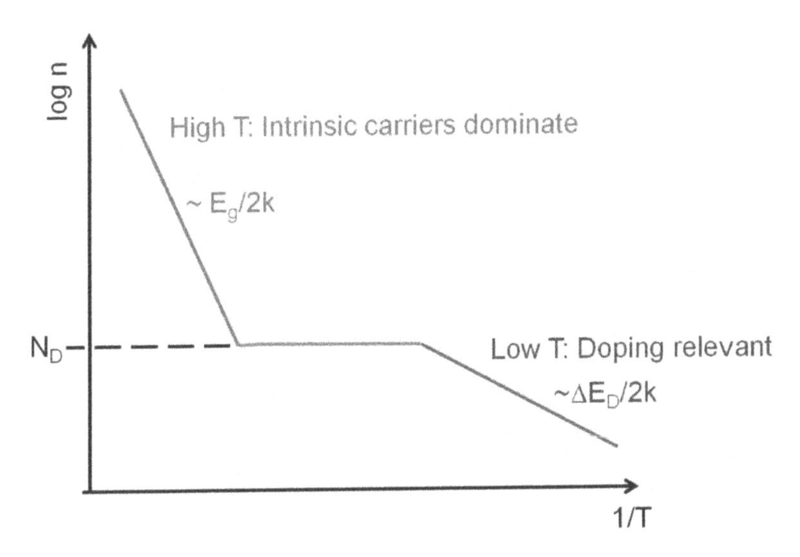

Figure 5.2 Temperature dependence of the carrier concentration in a doped semiconductor (schematically). At low temperatures, donors get more and more ionized with temperature. When all donors are ionized, the carrier concentration saturates at the donor concentration N_D up to those temperatures where the thermal excitation of carriers from the valence band into the conduction band starts to dominate.

extra electron of the donor is bound by an acceptor; hence, the resulting carrier concentration may be very low.

5.2 Carrier Mobility

The electrical conductivity of a material depends, besides the carrier concentration, on the characteristics of these carriers, in particular their **mobility**.

Remember that the current density j is given by

$$j = q \cdot n \cdot v_G, \tag{5.5}$$

with q = elementary charge.

The group velocity of the carriers v_G is created by and hence is proportional to the accelerating electric field E (which may be a consequence of an externally applied voltage):

$$v_G = \mu \cdot E. \tag{5.6}$$

The proportionality factor μ is called **carrier mobility**. Thus we get

$$j = q \cdot n \cdot \mu \cdot E.$$

How is the mobility related to the microscopic properties of the carriers?

In an electric field E, the carrier receives a force

$$F = m^* \frac{dv}{dt} = q \cdot E.$$

Hence $$\frac{dv}{dt} = \frac{q \cdot E}{m^*}.$$

After a short time τ, the carrier is scattered, i.e., its velocity is (in average) set back to zero, which can be described by

$$\frac{dv}{dt} = -\frac{v}{\tau}.$$

Thus we obtain

$$\mu = \frac{q\tau}{m^*}.$$

We observe that the mobility is

- proportional to the scattering time (the less scattering events, the better is the mobility) and
- inversely proportional to the effective mass (the heavier the carrier, the less mobile it is).

As discussed earlier, the carrier mass is determined by the band structure (curvature of dispersion relation) of the respective material.

On the other hand, **what determines the scattering time τ?**

There are different scattering mechanisms depending on temperature and other properties (e.g., impurity concentration; see Fig. 5.3):

- At **low** temperature, typically ionized impurity scattering dominates. Therefore, the mobility changes as follows:
    - $\mu_I \sim T^{3/2}$ (impurity scattering cross section is temperature dependent, mainly due to the temperature dependence of the thermal velocity of the carriers).

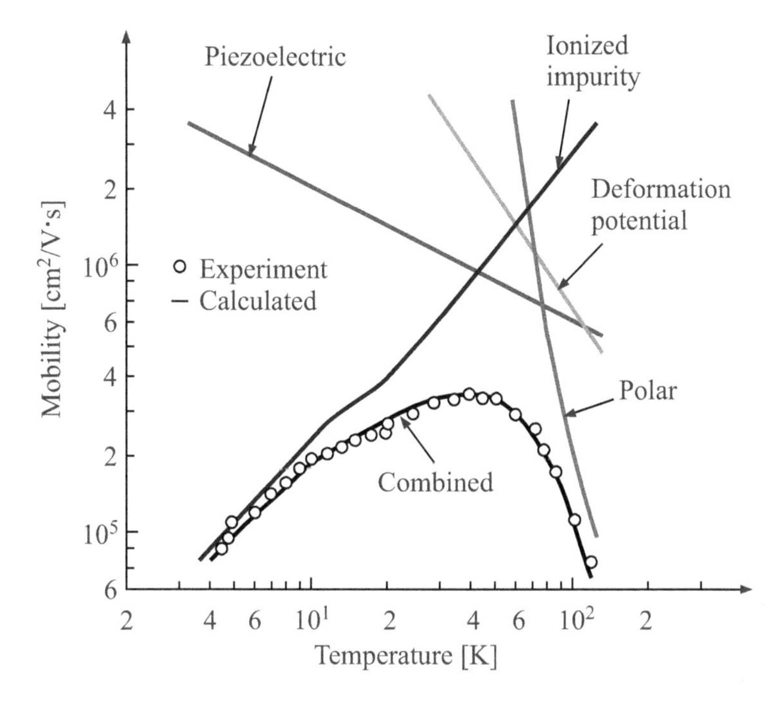

Figure 5.3 Electron mobility in bulk GaAs versus temperature (reprinted from [34], with the permission of AIP Publishing).

- $1/\mu_\mathrm{I} \sim N_\mathrm{D}, N_\mathrm{A}$.

  Neutral impurities are less important.

- At **mid** temperatures, different mechanisms may play a role:
  - Deformation potential scattering by acoustic phonons: $\mu_\mathrm{DP} \sim T^{-3/2}$. Phonons displace the atoms, i.e., they disturb the lattice periodicity and, hence, give rise to the scattering of the carrier wave (notice: no phonons available at $T = 0$).
  - Piezoelectrically active acoustic phonons: $\mu_\mathrm{PE} \sim T^{-1/2}$. The phonons deform the lattice thus generating a piezoelectric field which finally interacts with the charged carriers. Of course, this can only take place in lattices which contain two different atoms in the unit cell. See more about piezoelectricity in Section 11.6.

- At **high** temperatures, optical phonons (polar, PO or non-polar, NPO) dominate: $\mu_{\text{PO}} \sim T^{-2}$ and $\mu_{\text{NPO}} \sim T^{-3/2}$. Optical phonons are only present in crystals with two (or more) atoms in the unit cell, i.e., the two atoms oscillate anti-parallel. Polar optical phonons are only present, if the two atoms are different (normal case in compound semiconductors).

Additionally, alloy scattering may play a role in ternary and multinary crystals. The statistical distribution of the atoms disturbs the lattice periodicity.

These different scattering mechanisms contribute to the overall scattering time according to **Mattiesen's rule**:

$$\frac{1}{\tau} = \sum_i \frac{1}{\tau_i} \quad \text{or} \quad \frac{1}{\mu} = \sum_i \frac{1}{\mu_i} \tag{5.7}$$

Keep in mind that at low temperature, ionized impurities dominate, whereas at high temperatures, phonons dominate (as they are thermally generated).

5.3 Measurement of Electrical Properties

5.3.1 Hall Effect

Moving carriers in a magnetic field B feel the **Lorentz force** and thus are deflected. Hence, a **Hall** voltage arises in a sample perpendicular to the current direction. The Hall coefficient R is given by

$$R \equiv \frac{E}{j \cdot B} = \frac{r}{n \cdot q},$$

with j = current density, q = carrier charge, and E = electric field developing due to the carrier deflection (Hall field); r is a factor close to unity (i.e., $3\pi/8$).

Hence, this is a simple method to evaluate the carrier concentration n.

Additionally, the conductivity σ is measured. From $\sigma = \frac{j}{E} = q \cdot n \cdot \mu$, the mobility can be evaluated.

Typically, this experiment is carried out in a simple geometrical arrangement: In **van der Pauw geometry** [35], just a thin sample

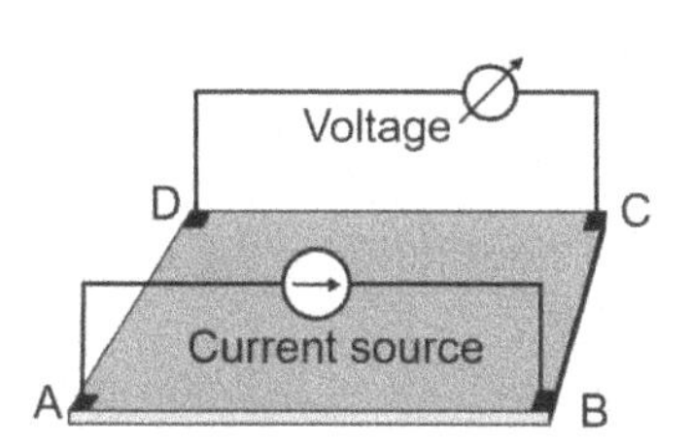

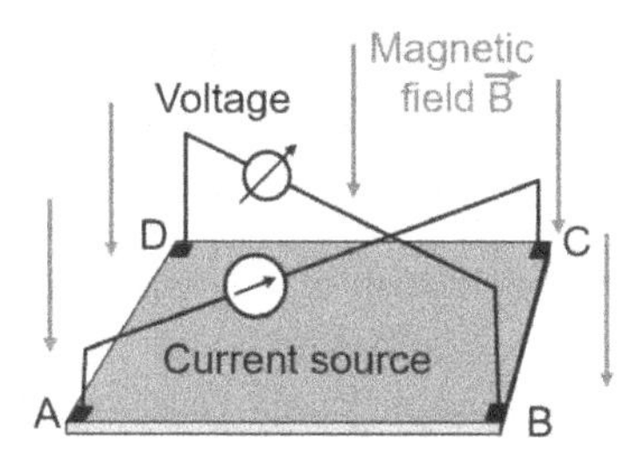

Figure 5.4 Hall measurement in van der Pauw geometry: In the first step, the conductance of the semiconductor sample is determined by measuring the resistances $R_{\mathrm{AB,CD}}$ and $R_{\mathrm{BC,DA}}$ (left). Then, the Hall voltage is obtained by measuring the resistance $R_{\mathrm{AC,BD}}$ with a magnetic field $\vec{B}$ applied perpendicular to the sample surface (right).

(layer) with a square shape area is taken with four contacts in the corners (Fig. 5.4).

The different "resistances" to be measured in this method (see Fig. 5.4) are defined as follows:

$$R_{\mathrm{ij,kl}} = \frac{V_{\mathrm{kl}}}{I_{\mathrm{ij}}},$$

with i, j, k, l being the contact points A, B, C, D in various combinations.

This is a four-point probe measurement, having the significant advantage that any contact resistance is compensated.

From these data, the sheet resistivity and the sheet carrier concentration can be determined. If the thickness d of the sample is known, then the specific resistivity ρ, the carrier concentration n and the mobility μ can be evaluated:

$$\rho = \frac{\pi d}{\ln 2} \cdot \frac{R_{\mathrm{AB,CD}} + R_{\mathrm{BC,DA}}}{2} \cdot f$$

$$n = \frac{2 \cdot B}{q \cdot d \cdot \Delta R_{\mathrm{AC,BD}}}$$

$$\mu = \frac{1}{q \cdot n \cdot \rho},$$

with f being a geometry factor which is close to unity for a square-shaped sample (see more details in [35]), q is the elementary charge and $\Delta R_{\mathrm{AC,BD}}$ is the difference of $R_{\mathrm{AC,BD}}$ measured with the magnetic field $\vec{B}$ applied perpendicular and inversely perpendicular to the sample ($\Delta R_{\mathrm{AC,BD}} = R_{\mathrm{AC,BD}}(B+) - R_{\mathrm{AC,BD}}(B-)$), whereas $B = |\vec{B}|$.

Notice that this method allows to determine directly the carrier type (whether electrons and holes) following from the sign of the Hall voltage (i.e., $R_{AC,BD}$).

In principle, this method can only determine the carrier concentration of the whole sample, i.e., it does not give the carrier profile in different layers of a heterostructure.

In simple cases (e.g., two layers), a respective evaluation may be possible, if some data are known from other measurements. Example: Data of layer 1 is known. Then measure n_{eff} and μ_{eff} of total structure (two layers). Evaluate properties of the second layer according to

$$n_{\text{eff}} = \frac{(n_1\mu_1 + n_2\mu_2)^2}{n_1\mu_1^2 + n_2\mu_2^2}$$

$$\mu_{\text{eff}} = \frac{n_1\mu_1^2 + n_2\mu_2^2}{n_1\mu_1 + n_2\mu_2}$$

Such an evaluation may be in particular necessary, if the layer to be studied is grown on a well-known conducting layer (e.g., conductive substrate).

By etching away one layer after another, this may be used to evaluate more complex structures (differential Hall measurement). However, this is rarely performed.

5.3.2 *C–V* Profiling

In many cases, we are interested in a carrier profile, i.e., in the information how the carriers are distributed along the growth direction of epitaxial layers or in a more complex multi-layer structure (e.g., in the many differently doped layers of a device). Then Hall measurements are not applicable. Alternatively, this can be done by measuring capacitance-voltage (C–V) curves.

Basic idea: In a Schottky contact (a non-Ohmic metal–semiconductor contact) with reverse bias, a depletion layer forms acting similar to the insulation layer in a capacitance (Fig. 5.5).[b] The properties of the depletion layer depend on the built-in voltage V_b, which is a direct consequence of the difference of the electron

[b]Basically the same happens in a reversely biased pn junction.

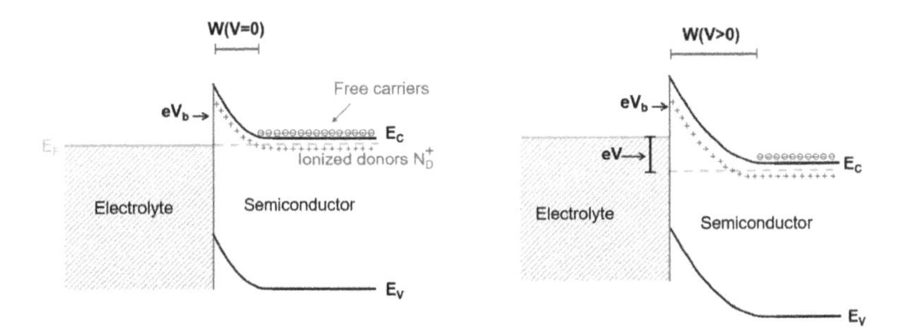

Figure 5.5 Schottky contact between a semiconductor and a perfect conductor in thermal equilibrium (left) and under reverse bias (right). The built-in voltage V_b depends on the respective materials. For C–V profiling, the conductor is formed by an electrolytic liquid, whereas in other cases, a metal may form a Schottky contact with a semiconductor. Notice the wider depletion region $W(V)$ in the right diagram.

affinity between the semiconductor and the metal (compare Fig. 2.4 and its explanation), the local dopant concentration N_D and the applied voltage V. When changing the applied voltage, then carriers flow into or out of the depletion layer, while the immobile dopants remain at their positions. For reverse bias, the bands get bent so that the electrons flow into the bulk semiconductor, leaving the ionized donors N_D^+ near the surface. Hence, the local charge density changes, i.e., this junction works indeed like a capacitance.

From Poisson's equation, which describes the relation between the local potential $\Phi(x)$ and the local charge density $\rho(x)$:

$$\frac{d^2\Phi(x)}{dx^2} = -\frac{\rho(x)}{\varepsilon_0\varepsilon_\mathrm{r}} = -\frac{qN_\mathrm{D}^+}{\varepsilon_0\varepsilon_\mathrm{r}} \tag{5.8}$$

we obtain after two integrations

$$\Phi(x) = -\frac{qN_\mathrm{D}^+}{2\varepsilon_0\varepsilon_\mathrm{r}}x^2\bigg|_0^W \tag{5.9}$$

In our case, we make a simple approximation where the depletion layer is completely depleted up to a depth W and then bulk properties are present. The applied voltage V and the built-in voltage V_b is just across this step-like depletion layer, i.e., $\Phi(0) = 0$, $\Phi(W) = V_\mathrm{b} + V$. Moreover, we assume homogeneous doping, i.e., a constant concentration of ionized donors N_D^+ over this distance. Then we can resolve Eq. 5.9 for the width W:

$$W = \left(\frac{2\varepsilon_0\varepsilon_r(V_b + V)}{qN_D^+}\right)^{1/2} = L_D\left(\frac{2q(V_b + V)}{kT}\right)^{1/2} \tag{5.10}$$

with the Debye length

$$L_D = \left(\frac{\varepsilon_0\varepsilon_r kT}{q^2 N_D^+}\right)^{1/2}$$

which describes the penetration depth of the electric field. For our problem, it defines the resolution limit: W must be larger than L_D. Typical values for the Debye length: $L_D \simeq 100$ nm for $n \simeq 10^{15}$ cm^{-3} and $L_D \simeq 5$ nm for $n \simeq 10^{10}$ cm^{-3}, respectively.

Now, we can obtain the total charge in the depletion region per area A:

$$\frac{Q}{A} = q \cdot N_D^+ \cdot W = q \cdot N_D^+ \cdot L_D\left(\frac{2q(V_b + V)}{kT}\right)^{1/2}$$

and hence the capacity per area as its derivative versus the applied voltage:

$$\frac{C}{A} = \frac{1}{A}\frac{dQ}{dV} = q \cdot N_D^+ \cdot \frac{dW}{dV} = \left(\frac{q\varepsilon_0\varepsilon_r N_D^+}{2(V_b + V)}\right)^{1/2}$$

V_b is normally not known beforehand. This can be resolved by evaluating additionally the derivative of the capacitance

$$\frac{1}{A}C' = \frac{1}{A}\frac{dC}{dV} = \left(\frac{q\varepsilon_0\varepsilon_r N_D^+}{8(V_b + V)^3}\right)^{1/2}.$$

Then we get finally N_D^+ at the position $x = W$:

$$N_D^+ = \frac{1}{\varepsilon_0\varepsilon_r A^2}\frac{C^3}{C'}. \tag{5.11}$$

How is the measurement performed?

- The depletion width W can be adjusted by a DC voltage V_0.
- We then can measure C by evaluating the flowing charge ΔQ by modulating V_0 with a small AC voltage $\Delta V_1 \sin\omega_1 t$.
- Similarly, $C' = \frac{dC}{dV}$ can be obtained by modulating C with even smaller voltage $\Delta V_2 \sin\omega_2 t$ with larger frequency $\omega_2 > \omega_1$.

However, only a limited penetration depth (limited by the breakdown voltage, which is inversely proportional to the carrier concentration) can be evaluated by this method. The larger the carrier concentration, the smaller is the penetration depth W.

This limitation can be overcome by the method of C–V **profiling**:

Now, the Schottky contact is formed by an electrolytic liquid which also can act as an etchant to the semiconductor. Hence, the sample is etched during the measurement, and a carrier concentration profile can be measured.

As this is an electrochemical etching process, the amount of etched material is directly correlated to the amount of flown electrical charge. Hence, the etching depth W_E can be quantitatively evaluated from the flown charge:

$$W_E = \frac{M}{z \cdot F \cdot \rho \cdot A} \int_0^t I dt,$$

with z = number of carriers per molecule, F = Faraday constant (96496 C), M = molecular weight, ρ = mass density and A = area of etched crater. In some cases, the etching is further triggered by light ("photo-electrochemical etching").

An example for a measurement cell is displayed in Fig. 5.6.

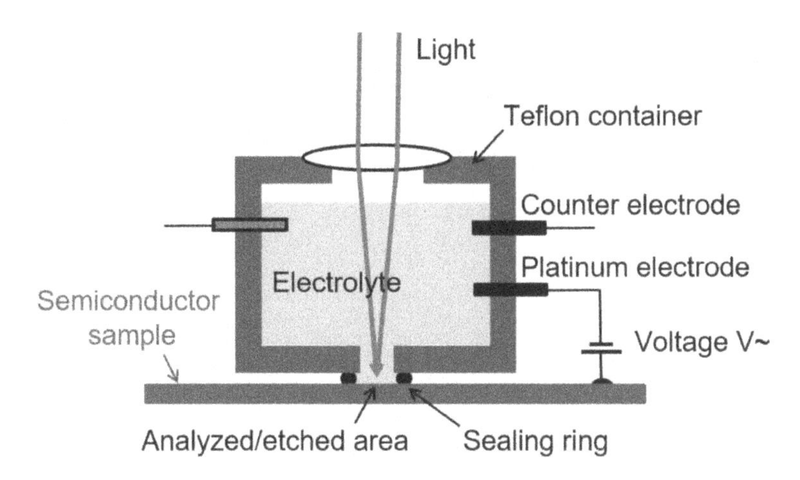

Figure 5.6 C–V-profilometer measurement cell. The counter electrode supplies the current for etching, the Pt electrode is used for measurement. The calomel electrode serves as a reference.

A typical etchant liquid is ammonium tartrate/NH_3 (used for GaAs or GaInP) or KOH (for GaN).

In such an experiment, the procedure is periodically switched between electrochemical etching and C–V measurement. Typical frequencies for the latter are $\omega_1 \simeq 3$ kHz and $\omega_2 \simeq 40$ kHz.

Via L_D, the depth resolution depends on doping.

Disadvantage of this method: The sample is destroyed during the measurement; hence, it can only be checked by a second measurement at another position.

Advantage: This method can be applied to quite complex device structures (e.g., laser structures), where the doping may vary from layer to layer (see Fig. 5.7).

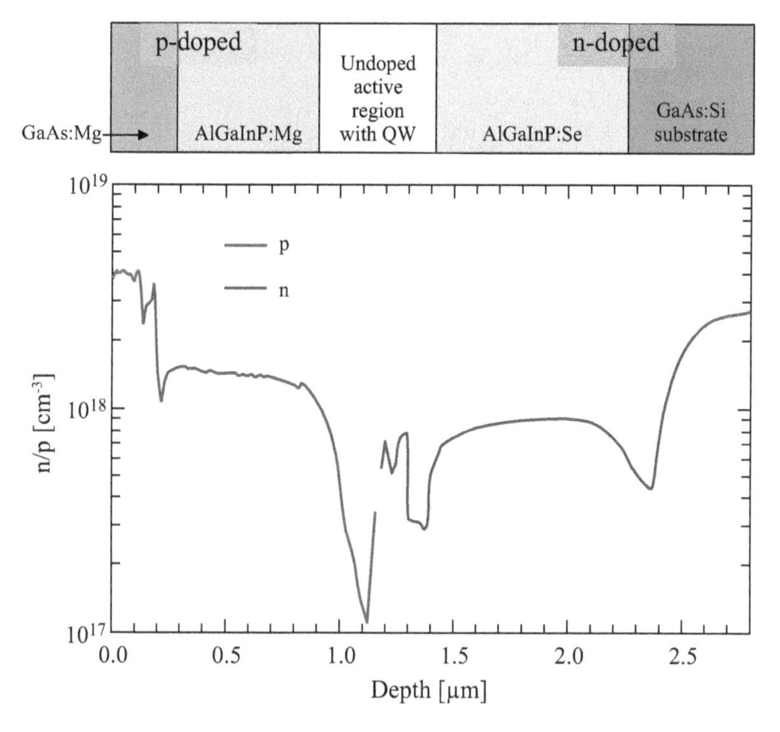

Figure 5.7 C–V doping profile of an AlGaInP laser diode structure (after [36]).

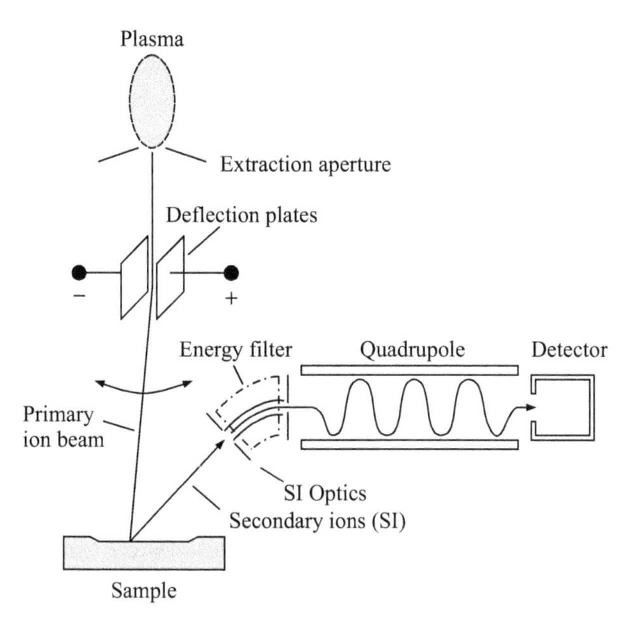

Figure 5.8 Secondary ion mass spectrometry machine (schematically, after [39]).

5.4 Secondary Ion Mass Spectrometry

Besides the *carrier* concentration profile, very often the *dopant* concentration profile should be known. This can be done by secondary ion mass spectrometry (SIMS). For this technique, the reader may find more details in [37, 38].

The basic idea of this method is: Remove the surface of a specimen layer by layer by ion bombardment, analyze quantity and type of removed atoms (Fig. 5.8).

Following are the main components of a SIMS machine:

- Ion source providing high-energy ions for the sputtering process:
  Two primary ions (Cs or O) are mainly used. With Cs, a dramatic increase of ion yield is obtained for *electroneg-ative* elements, whereas O results in higher yield for less

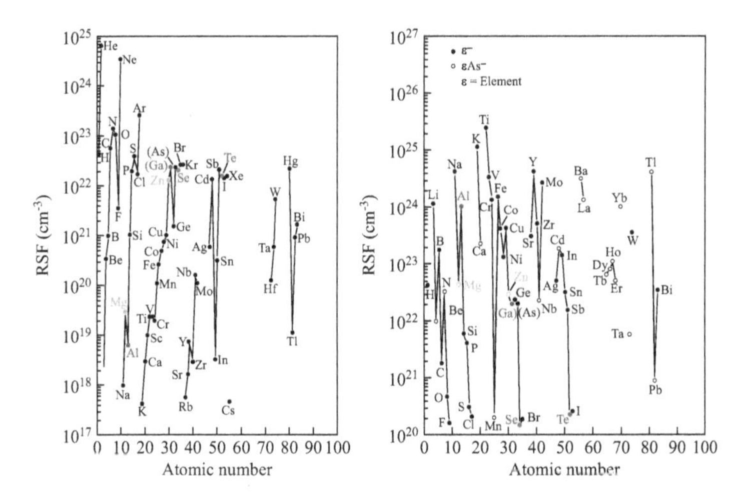

Figure 5.9 Relative sensitivity factors (RSF) of some secondary ions after sputtering GaAs with an O_2^+ (left) or Cs^+ (right) primary ion beam (after [37]).

electronegative elements (Fig. 5.9). Typical ion voltages range between 3 and 15 kV.

- Electrostatic deflection stage for primary beam:
  Necessary to scan the primary ion beam over the surface. Depending on the problem, *gating* may be applied, i.e., the secondary ions are only analyzed in those times when the primary beam is not near the crater walls in order to suppress parasitic signals of the walls. This gating can also be applied after the measurement, if all data have been stored together with the information about their local generation.
- Sample to be evaluated:
  In the sample, the depth resolution is limited by the finite penetration depth of the primary beam (depending on the ion energy). Due to the high energy of the primary ions, some atoms of the layer to be analyzed may be pushed deeper into the material. Hence, they will be assigned erroneously later to this deeper layer, which effectively reduces the depth resolution (Fig. 5.10). Today, a depth resolution down to about 1–2 nm can be achieved. The concentration resolution limit of the atoms to be analyzed depends on the elements. Typical values go down to 10^{15}–10^{16} cm^{-3} and below.

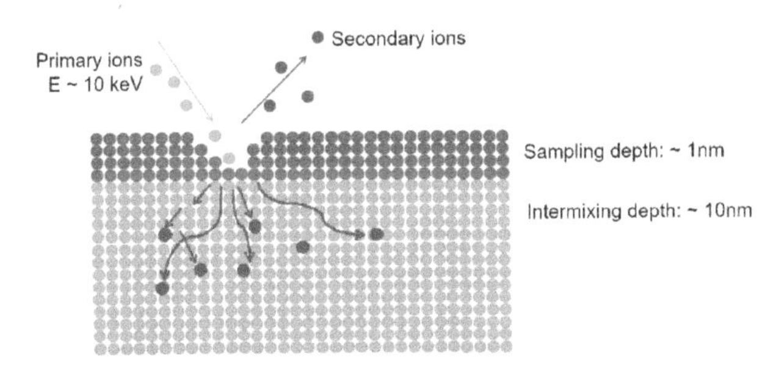

Figure 5.10 Intermixing induced by the primary ion beam (schematically).

By the above-mentioned gating, also some lateral resolution of a few micrometers can be obtained.

- Energy filter:
  It is used to reject neutral and high-energy ions. It performs a pre-selection of ions regarding their energy before they enter the mass spectrometer. In some spectrometers, it is just a part of it. In systems with a quadrupole mass spectrometer, it is a separate device.
- Quadrupole filter (Fig. 5.11):

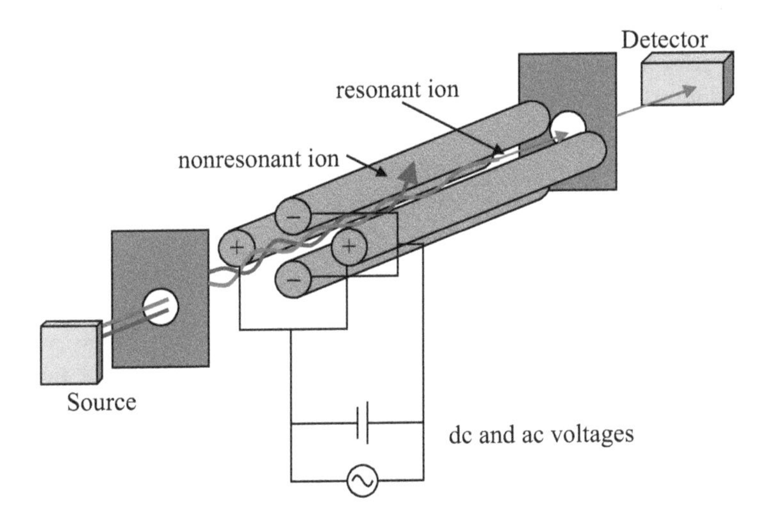

Figure 5.11 Quadrupole mass filter (schematically, from [40]).

By adjusting two voltages, it can be realized that only ions of one specific e/m ratio can travel through the filter. Compared to other e/m filters, it has the following advantages:

  - High resolution for e/m between 3 and 400; easy separation of isotopes
  - high scan speeds for $m = 3 - 250$
  - high precision when moving from mass to mass (no magnetic field involved)
  - high sensitivity (ppb)
  - high reliability

  Note that only a specific e/m ratio can be resolved. This may lead to misinterpretations if molecular ions with the same e/m are present (e.g., N_2 has the same molecular mass of 28 as a single Si ion).

- Ion detector: Works similar to a photo-multiplier (see Fig. 6.10).

The number of secondary ions depends—besides the primary beam details—on the host material type (matrix-dependent binding energy and ionization probability, called "matrix effect"). This makes the determination of absolute concentrations difficult. This problem is solved by using reference calibration samples with well-known impurity or doping concentration. They are fabricated by ion-implantation which enables a precise prediction of the implantation profile depending on the ion beam current and the ion energy.

Figure 5.12 shows the SIMS data measured on a GaN-based laser test structure where such quantitative evaluation of the data was performed.

Other problems:

- The charging of insulating samples by the primary ions may lead to its eventual reflection. This can be managed by a simultaneous bombardment with electrons.
- Intermixing of the material by the primary ion beam leading to a low depth resolution as discussed above (Fig. 5.10). Can be solved by lower ion beam energies and/or by glancing incidence of the primary beam (e.g., 70° instead of 0°).

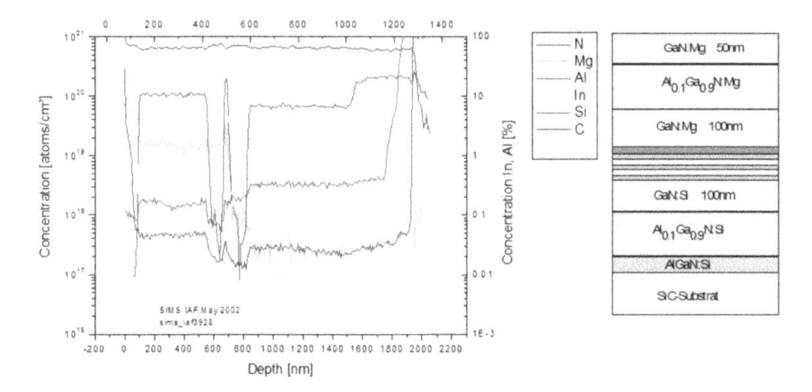

Figure 5.12 SIMS data (left) of a GaN-GaInN laser test structure (layer sequence shown right). These data were measured by M. Mayer, IAF Freiburg.

Problems

(1) Deduce the relation how the Fermi level position depends on the carrier concentration.

(2) Explain qualitatively, how the carrier concentration in a doped semiconductor changes with temperature.

(3) Explain how the donor ionization energy can be determined.

(4) What does "compensation" mean with respect to doping and impurities? How large is the carrier concentration (qualitatively) in a compensated semiconductor?

(5) Discuss some advantages and limitations of the van der Pauw–Hall experiment.

(6) How can you determine which types of carriers are present in an epitaxial layer?

(7) How is the carrier mobility related to microscopic properties of a semiconductor?

(8) How can we measure the carrier mobility of a semiconductor layer?

(9) C–V profiling:

 (a) Explain what can be determined by C–V measurements. How does it work?

(b) How can the depth be determined in C–V profiling experiments?

(10) Explain briefly: How does SIMS work? What can be measured?

Chapter 6

Optical Processes in Semiconductors: Optical Spectroscopy

Again, only a brief introduction to this topic can be made in this book. The interested reader can find more about these topics in many excellent textbooks, e.g., [4, 41, 42].

6.1 Basic Optical Processes

Regarding the interaction of semiconductors and light, three basic processes require special focus:

- **Absorption**: An electron is lifted from the valence band (state with energy $E_1 < E_v$) into the conduction band (state with energy $E_2 > E_c$) by absorbing a photon of the respective energy $h\nu = E_2 - E_1$ (Fig. 1.16 left). The optical transition rate $g_{vc}(\vec{k})$ depends on the number of electrons in the valence band $f_v(E_v)$, the number of open states in the conduction band $[1 - f_c(E_c)]$ and the photon density $\rho_{ph}(h\nu)$:

$$g_{vc}(\vec{k}) = B_{vc} \cdot f_v(E_v) \cdot [1 - f_c(E_c)] \cdot \rho_{ph}(h\nu), \tag{6.1}$$

Compound Semiconductors: Physics, Technology, and Device Concepts
Ferdinand Scholz

ISBN 978-981-4774-07-9 (Hardcover), 978-1-315-22931-7 (eBook)
www.panstanford.com

where B_{vc} is the transition rate per photon depending on the band structure. In fact, this is the respective Einstein coefficient (c.f. black body radiation).

- **Stimulated emission**: An electron in the conduction band falls back to the valence band, triggered by a photon of respective energy (Fig. 1.16, center). Now, occupied states in the conduction band $f_c(E_c)$ (i.e., excited carriers) and empty states in the valence band $[1 - f_v(E_v)]$ are required together with photons $\rho_{ph}(h\nu)$:

$$g_{cv}^{stim}(\vec{k}) = B_{cv} \cdot f_c(E_c) \cdot [1 - f_v(E_v)] \cdot \rho_{ph}(h\nu) \qquad (6.2)$$

  By this process, another photon is generated which is absolutely identical to the incoming photon regarding all its quantum physical properties (e.g., wavelength, momentum, polarization). Therefore, light is amplified in this process. This is the key process for any laser action (see Section 12.2). As originally postulated by Einstein, this is indeed the inverse process to absorption.

- **Spontaneous emission**: After an average life time τ_s, the excited electron in the conduction band falls back into the valence band spontaneously (if it was not triggered by stimulated emission) to bring back the system into thermodynamic equilibrium. In case of radiative recombination, a photon of the respective energy is emitted (Fig. 1.16 right). The transition probability reads

$$g_{cv}^{spont}(\vec{k}) = A_{cv} \cdot f_c(E_c) \cdot [1 - f_v(E_v)]. \qquad (6.3)$$

In the thermodynamic equilibrium, the net rate should be zero, i.e.,

$$g_{vc} = g_{cv}^{stim} + g_{cv}^{spont}.$$

As absorption and stimulated emission are just opposite processes, it follows that $B_{cv} = B_{vc}$ (can be proved by perturbation theory). Moreover, the following relation combines spontaneous and stimulated emission [43]:

$$A_{cv} = \frac{8\pi \bar{n}^3 (\Delta E)^2}{h^3 c^3} B_{cv} = \frac{1}{\tau_r}, \qquad (6.4)$$

with τ_r being the spontaneous radiative life time, $\bar{n}$ = refractive index, $\Delta E = h\nu$ = energy difference of two respective states which is equal to the energy of the respective photons, h = Planck's constant, c = speed of light.

Please notice: The larger ΔE ($\sim E_g$), the smaller τ_r; hence, the faster the system returns to equilibrium after excitation. Therefore, it becomes more and more difficult to realize large excitation densities or population inversion, which is a prerequisite for laser action.

B_{cv} can be calculated from **Fermi's golden rule** by applying perturbation theory:

$$B_{cv} = \frac{\pi}{2\hbar} \left| \langle \Psi_{c,k} | W | \Psi_{v,k} \rangle \right|^2 \tag{6.5}$$

W is in general the perturbation operator. In our case (optical transition), it is effectively the electric field of the photon $[-q\vec{E}\hat{r}\cos(\vec{k}\vec{r} - \omega t)]$.

These formulas hold even in case of non-thermodynamic equilibrium! Then the Fermi distribution has to be replaced by the quasi-Fermi distributions f_c and f_v (see below).

6.2 Steady-State Excitation

In thermodynamic equilibrium, the mass action law (c.f. Eq. 1.42) is valid: $n_0 \cdot p_0 = n_i^2(T)$. Notice: n_0 and p_0 may be changed by doping.

Of course, in thermodynamic equilibrium, no net change of the carrier distribution occurs.

This equilibrium may be perturbed by excitation:

- illumination
- electrical current
- others (e.g., electron beam, chemical excitation)

After excitation, we may find many electrons in the conduction band and many holes in the valence band. Before they recombine, they may find their respective thermodynamic equilibrium in either band by fast intra-band relaxation processes. Then, we can define the quasi-Fermi levels E_{Fn} and E_{Fp}.

Determination of Quasi-Fermi levels: The spontaneous recombination rates R_i of holes and electrons, respectively, are given by

$$R_\mathrm{n} = \frac{n}{\tau_\mathrm{n}} \quad \text{and} \quad R_\mathrm{p} = \frac{p}{\tau_\mathrm{p}},$$

with τ_i being the spontaneous life times of electrons and holes, respectively. On the other hand, carriers in either band are generated by the generation rates G_n and G_p, respectively.

In a steady-state situation (i.e., continuous excitation) the generation and recombination rates should be equal, i.e.,

$$n = G_\mathrm{n} \cdot \tau_\mathrm{n} \quad \text{and} \quad p = G_\mathrm{p} \cdot \tau_\mathrm{p}, \tag{6.6}$$

From Eq. 1.36

$$n(T) = N_\mathrm{C} \cdot e^{\frac{E_\mathrm{F} - E_\mathrm{C}}{kT}}$$

the Fermi level can be deduced:

$$E_\mathrm{F} = E_\mathrm{c} - kT \ln\left(\frac{N_\mathrm{c}}{n}\right)$$

Similarly we get for the holes:

$$E_\mathrm{F} = E_\mathrm{v} + kT \ln\left(\frac{N_\mathrm{v}}{p}\right)$$

In our case of a steady-state excited semiconductor (i.e., out of thermal equilibrium, but in a new equilibrium state stabilized by the steady-state excitation), we take the carrier concentrations from Eq. 6.6 by just replacing the Fermi level by the quasi-Fermi levels:

$$E_\mathrm{Fn} = E_\mathrm{c} - kT \ln\left(\frac{N_\mathrm{c}}{G_\mathrm{n}\tau_\mathrm{n}}\right) \quad \text{and}$$

$$E_\mathrm{Fp} = E_\mathrm{v} + kT \ln\left(\frac{N_\mathrm{v}}{G_\mathrm{p}\tau_\mathrm{p}}\right).$$

Thus, the separation of the two quasi-Fermi levels can be determined:

$$E_\mathrm{Fn} - E_\mathrm{Fp} = E_\mathrm{c} - E_\mathrm{v} + kT \ln\left[\frac{G_\mathrm{n}\tau_\mathrm{n}}{N_\mathrm{c}} \cdot \frac{G_\mathrm{p}\tau_\mathrm{p}}{N_\mathrm{v}}\right] \tag{6.7}$$

This has big importance in particular for semiconductor laser diodes (see later). It describes nicely the carrier distribution situation in a semiconductor, where some excitation is sustained by external means (it may be optical excitation or pumping by electrical current, etc.) and hence, whether and how far a system is out of thermodynamic equilibrium.

Notice: The product of electron and hole concentration now reads

$$n \cdot p = n_\mathrm{i}^2 \cdot e^{\frac{E_\mathrm{Fn}-E_\mathrm{Fp}}{kT}},$$

i.e., the mass action law (Eq. 1.42) is not valid under non-equilibrium conditions where $E_\mathrm{Fp} \neq E_\mathrm{Fn}$. However, the number of (excited) carriers increases drastically for increasing separation of the quasi-Fermi levels.

There are different spontaneous recombination mechanisms which will bring back the system into thermodynamic equilibrium:

- Radiative recombination: Interaction with electromagnetic field of photon. As mentioned above, for spontaneous recombination, electron and hole must interact. Hence, the spontaneous recombination rate can be determined in a somewhat simplified way from Eq. 6.3 as

$$R_\mathrm{spontan} = B \cdot n \cdot p \tag{6.8}$$

  with B = so-called bimolecular recombination coefficient. For the case of strong excitation ($\Delta n = \Delta p \gg n, p$) we thus find:

$$R_\mathrm{spontan} = B \cdot \Delta n^2$$

- Non-radiative recombination:
  - **Shockley–Read–Hall recombination**: Electrons and holes get trapped (independently) at some impurities or deep levels. They may lose their energy to, e.g., phonons in several non-radiative steps. Hence, this process is governed by the **minority carrier concentration**:

$$\text{for } n \gg p: \quad R_\mathrm{p} = \frac{p_\mathrm{n} - p_\mathrm{no}}{\tau_\mathrm{p}} = \frac{\Delta p}{\tau_\mathrm{p}}$$

$$\text{for } n \ll p: \quad R_\mathrm{n} = \frac{n_\mathrm{p} - n_\mathrm{po}}{\tau_\mathrm{n}} = \frac{\Delta n}{\tau_\mathrm{n}},$$

    with Δn, Δp = number of excited minority carriers.

- **Auger recombination**: Similar as for atomic levels (see in situ characterization methods in MBE), an Auger process can also take place in a semiconductor. Example: An electron from the conduction band recombines with a hole in the valence band. The extra energy may be provided to another electron in the valence band, which hence is excited to another state (see Fig. 6.1) from where it typically relaxes non-radiatively. Now three particles are involved; thus the recombination rate reads

  $$R_{\text{Auger}} = (C_\text{n} \cdot n + C_\text{p} \cdot p)(np - n_\text{i}^2),$$

  with C_i = Auger coefficients for Auger electrons or holes, respectively. For $\Delta n \gg n, p$ this results in

  $$R_{\text{Auger}} = C_\text{n} \cdot \Delta n^3.$$

Finally, the total recombination rate is described by

$$\frac{d\Delta n}{dt} = -R(\Delta n) \quad \text{with} \quad R(n) = A \cdot \Delta n + B \cdot \Delta n^2 + C \cdot \Delta n^3. \quad (6.9)$$

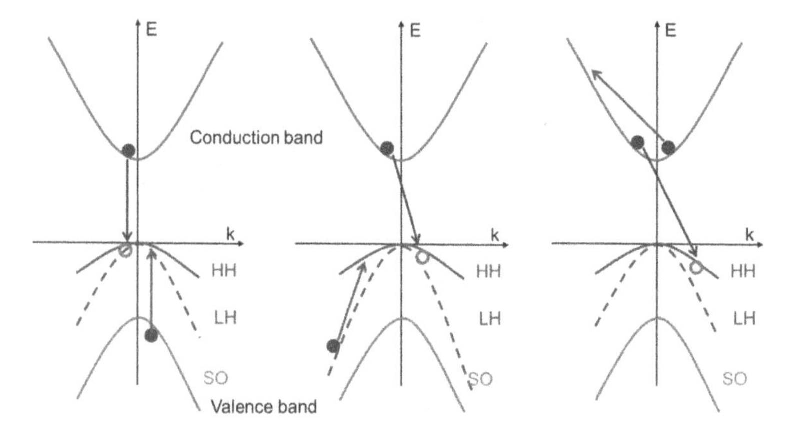

Figure 6.1 Different possibilities for a recombination process involving an Auger effect. Left: The energy of the recombining electron is transferred to an electron in the split-off valence band. This Auger process is typically important in semiconductors where the band gap is similar to the energy difference of the split-off band and the HH/LH bands, hence in materials with small band gap. Then, this effect is significantly in competition with laser-relevant radiative transitions. Middle and right: Other possibilities where the third carrier is excited in the valence or in the conduction band. Notice that as usual, both energy and momentum conservation need to be fulfilled.

This equation is often referred as *ABC model*. By measuring the PL quantum efficiency $\eta_{\mathrm{QE,\,PL}}$, i.e., the relation between excitation power $I_{\mathrm{excitation}}$ and photoluminescence I_{PL}, we thus may determine which process dominates.

Under steady-state conditions, the excitation rate for carriers equals the total recombination rate $R(\Delta n)$, whereas the radiative recombination rate is given by the term $B \cdot \Delta n^2$. Hence, the efficiency is given by

$$\eta_{\mathrm{QE,\,PL}} = \frac{\text{Number of detected photons}}{\text{Number of exciting photons}} = \frac{B \cdot \Delta n^2}{A \cdot \Delta n + B \cdot \Delta n^2 + C \cdot \Delta n^3}, \quad (6.10)$$

(c.f. Section 12.1.2) leading to an increasing efficiency at low excitation powers where the term $A \cdot \Delta n$, while it decreases again at high powers for $C \cdot \Delta n^3$ dominating (Fig. 6.2).

Looking closer, we can identify more specific recombination channels like

- Radiative recombination via shallow donors or acceptors: A carrier may be first trapped by a donor or acceptor before it finally recombines (conduction band—acceptor or donor—valence band transition). For high levels of both types of impurities, donor–acceptor recombination can be observed. The Coulomb energy of interaction between two ions lowers the energy by $e^2/4\pi\varepsilon_0\varepsilon_{\mathrm{r}}R$ where R describes the distance (and hence the concentration) of the impurities and $\varepsilon_0\varepsilon_{\mathrm{r}}$ is the dielectric constant of the material:

  $$h\nu = E_{\mathrm{g}} - (E_{\mathrm{D}} + E_{\mathrm{A}}) + e^2/4\pi\varepsilon_0\varepsilon_{\mathrm{r}}R$$

  The donor activation energy E_{D} can be obtained from the hydrogen analogy model (see Section 5.1).
  Therefore the pair band transition D_0-A_0 can be used to identify the nature of the acceptor.
  This D_0-A_0-transition, visible at low temperature (liquid He), changes into an e-A_0-transition at higher temperature due to thermal excitation of the donors (with the smaller activation energy).

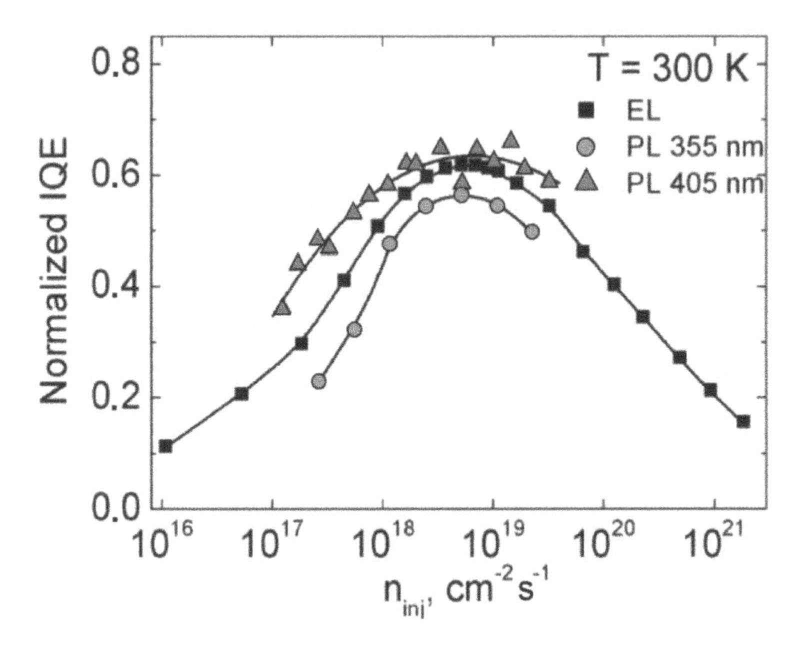

Figure 6.2 Photoluminescence (PL) and electroluminescence (EL) efficiency (intensity divided by excitation power) versus excitation power (in units of excited carriers per time) measured on a commercial GaN-based blue LED-structure containing $In_{0.15}Ga_{0.85}N/GaN$ multi quantum wells and barriers with thicknesses of 2.5 and 15 nm, respectively (from [44]).

- Exciton recombination. If (e.g., after excitation) both electrons and holes are present, an electron and a hole may interact via their Coulomb attraction forming a so-called **exciton**, again very similar to the formation of a hydrogen atom from a proton and an electron. By this formation process, the total energy of the pair is slightly lowered. If the two particles recombine anyway after a while, they emit a photon which is slightly lower than the band gap in energy.

  In semiconductors, the distance between the electron and hole is larger than the distance of the atoms (typically some 10 nm). Therefore, the exciton already "sees" the crystal potential averaged as continuum (Wannier exciton, very similar as discussed above for the dopants). Again, the crystal properties are included in the dielectric constant ε_r of the semiconductor, whereas all other parameters are very similar as for the hydrogen atom (typically in vacuum, i.e., $\varepsilon_{\mathrm{r}} = 1$), resulting again in a hydrogen-like relation:

$$E_{\mathrm{n}} = \frac{m_{\mathrm{r}} e^4}{8h^2(\varepsilon_0\varepsilon_{\mathrm{r}})^2 n^2} \quad \text{with} \quad \frac{1}{m_{\mathrm{r}}} = \frac{1}{m_e^*} + \frac{1}{m_h^*} \quad \text{reduced mass.}$$

Typical binding energies (according to this formula) describing the energy to separate the exciton, i.e., to bring back the carriers into the conduction and valence band, respectively, are in the range of 3–4 meV for InP and GaAs and 26 meV for GaN. Additionally, they may be (weakly) bound to some donor or acceptor impurity states lowering their energy further. In many cases, their fingerprint in spectroscopy dominates the characteristic spectra of high-purity semiconductors at low temperatures. They produce very sharp photoluminescence lines and thus allow to determine the nature of acceptors and even donors by resolving the small differences in E_{D}, as their position is directly related to the donor or acceptor activation energy (Hayne's rule).

These different radiative recombination processes compete with each other and with non-radiative recombination via multi-phonon processes or defects and impurities. All these processes have their characteristic time constants τ_i.

The total carrier lifetime is then given by

$$\frac{1}{\tau_{\text{total}}} = \sum \frac{1}{\tau_{\text{radiative}}} + \sum \frac{1}{\tau_{\text{nonradiative}}}. \tag{6.11}$$

As a consequence of the direct band structure of most compound semiconductors, radiative recombination dominates in a wide range of carrier concentrations. The spontaneous radiative recombination times depend strongly on material and structures. Typical values are in the nano- to microsecond range. However, for very small and very large carrier concentrations, i.e., excitation, non-radiative processes start dominating as nicely described by the *ABC model* (Eq. 6.9). This is of particular importance for GaN-based LEDs the efficiency of which decreases for larger current densities (see Section 12.1.2).

6.3 Optical Characterization Methods

Owing to the numerous interaction mechanisms of light and matter as briefly discussed above, optical (spectroscopic) characterization is a powerful tool to learn many details about the respective semiconductor. It is (in most cases) non-destructive and typically does not need complicated sample preparation. Hence, it can be applied comparably simply.

Which characteristic data can be measured?

- band gap (which may be translated into composition of ternary layers)
- material quality, e.g., composition uniformity, strain, defects, etc.
- impurities, doping characteristics
- properties of hetero structures (e.g., quantum wells, see later)
- bulk and surface information (depending on measurement details)
- carrier dynamics (e.g., recombination times, etc.)
- more details of the band structure

The term "optical spectroscopy" comprises many different techniques which can be applied to compound semiconductors, differing in the way how the sample is excited and/or how the signal is acquired. Some are mentioned in the following:

Excitation of a semiconductor sample with light may lead to the following:

- reflection of light

  → optical microscope
  → ellipsometry
  → reflection spectroscopy

- emission of light

  → photoluminescence
  → gain spectroscopy
  → Raman spectroscopy

- transmission of light
  - → absorption coefficient
  - → infrared spectroscopy
- absorption of light changing other material properties
  - → photoconductivity
  - → photoelectron spectroscopy
  - → photothermal deflection spectroscopy

Moreover, the sample may be excited by other means leading to light emission which can be spectroscopically analyzed:

- electrical current → electroluminescence (LED)
- electron beam (e.g., in an electron microscope) → cathodoluminescence
- chemical energy → chemoluminescence

Only a few of these many methods are discussed in detail hereafter.

6.3.1 Photoluminescence

Basically, luminescence means: A material (or sample) is excited, i.e., carriers are lifted to higher energy states by externally applied excitation. Then, they will fall back to their equilibrium state. As discussed above, this may be accompanied by the emission of a photon which fulfills the conservation of energy. This spontaneously emitted light is called **luminescence**. Depending on the way of excitation (e.g., light, electron beam, electrical current), the process is called photo-, cathodo-, or electro-luminescence, respectively.

One of the simplest methods to analyze some material is **photoluminescence** (PL): Excite the sample by shining some light on it, then detect the emitted light, analyze it with respect to wavelength (Fig. 6.3). If necessary, more details can be analyzed, e.g., decay time, polarization, local origin, etc.

In photoluminescence, the exciting photon energy must be higher than the energy state of the recombination path to be studied. Typically, it is larger than the band gap of the semiconductor leading to a strong absorption and to a high density of excited carriers far

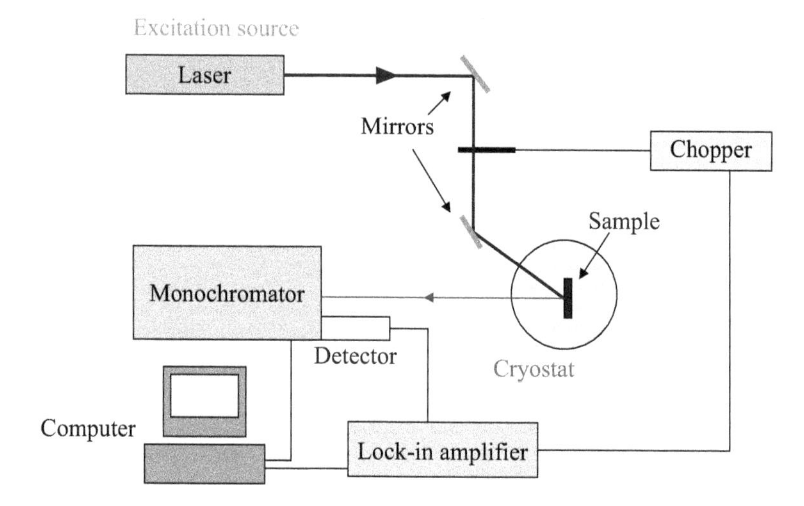

Figure 6.3 Photoluminescence setup (schematically). Typically, the light source is a laser as it provides fairly high excitation power. To act as an excitation source, the laser photon energy should be larger than the band gap of the semiconductor under investigation. The emitted light is spectrally resolved in the monochromator and finally detected in the detector. The lock-in amplifier (together with the chopper) helps to increase the signal-to-noise ratio and to suppress any unwanted background light.

above the band edge. Before the final radiative transition occurs, these excited carriers typically first relax very fast to lower states in the same band. As discussed above, such states may be the band edge, excitonic levels, impurity levels, etc. Hence, the luminescence signal does not reflect directly the excitation process, but the final recombination process thus providing detailed information of the material itself (Fig. 6.4). There is no direct relation between the exciting photon and the finally emitted photon (in terms of wavelength, phase, time).

The different recombination paths and the respective carrier redistribution after relaxation typically change their properties with temperature. At low temperature, the lowest transition energies, e.g., excitonic transitions (donor (D^0, X)- or acceptor (A^0, X)-bound excitons, free exciton FE) or donor–acceptor transitions (D^0, A^0) (see Fig. 6.4) can be observed, whereas at higher temperatures (e.g., room temperature) direct band to band recombination may

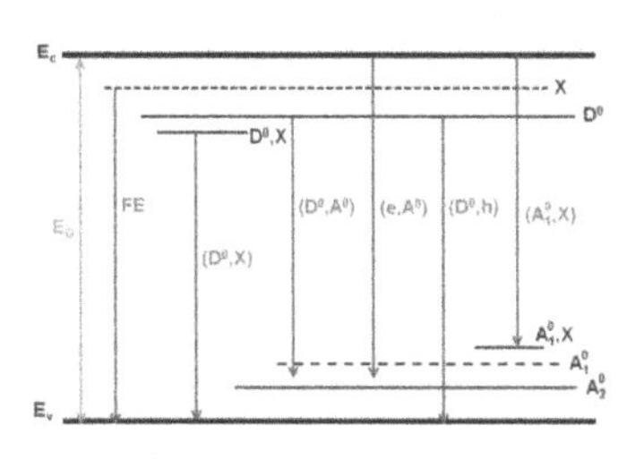

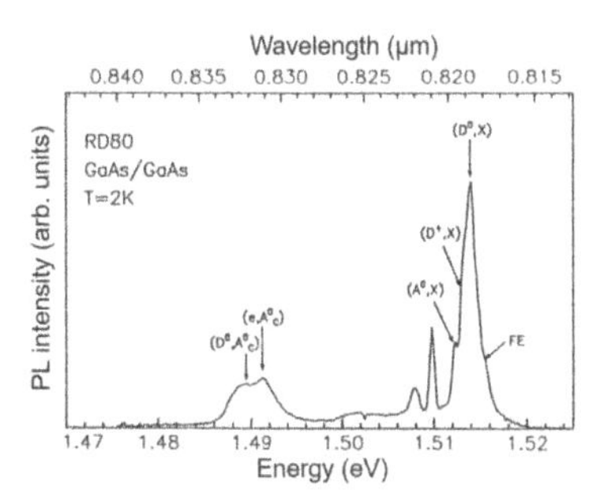

Figure 6.4 Left: Different recombination paths of an excited carrier in direct compound semiconductors (schematically). Right: Respective photoluminescence spectrum of GaAs around the band gap region taken at low temperature. From [45].

be dominant due to thermal population of the band edges and the much higher recombination probability of this process. Moreover, thermal broadening may occur due to the increased interaction with phonons.

Time-resolved studies of specific spectral lines will provide direct access to the respective relaxation or recombination paths.

6.3.2 Absorption

Light traveling through some material is more or less weakened by absorption. For a very thin layer, the change of intensity ΔI is proportional to the intensity I and to the traveling distance Δx (**Lambert's law**). The respective proportionality factor is called **absorption coefficient** α:

$$\Delta I = -\alpha \cdot I \cdot \Delta x \tag{6.12}$$

This can be translated into a differential equation

$$\frac{dI}{dx} = -\alpha \cdot I \quad \text{with the solution} \quad I(x) = I(0)e^{-\alpha x}. \tag{6.13}$$

The absorption coefficient is a material related coefficient. It is typically a function of wavelength: $\alpha = \alpha(\lambda)$.

It is a part of the complex refractive index

$$\tilde{n} = \bar{n} + i\kappa \quad \text{with} \quad \alpha = \frac{4\pi}{\lambda}\kappa = \frac{2\omega}{c}\kappa,$$

where $\bar{n}$ is the real refractive index and c = speed of light.

Moreover, notice the relation between the complex refractive index and the complex dielectric constant:

$$\tilde{\varepsilon} = \tilde{n}^2 = \varepsilon_1 + i\varepsilon_2,$$

$$\text{with} \quad \varepsilon_1 = \bar{n}^2 - \kappa^2 \quad \text{and} \quad \varepsilon_2 = 2\bar{n}\kappa = \frac{\bar{n}\alpha\lambda}{2\pi}.$$

Real and imaginary parts of the dielectric constant, i.e., refractive index $\bar{n}$ and absorption coefficient α, are linked by the **Kramers–Kronig relation**:

$$\bar{n}(\omega) = 1 + \frac{c}{\pi} \cdot \mathrm{CH} \int_0^\infty \frac{\alpha(\Omega)}{\Omega^2 - \omega^2} d\Omega, \qquad (6.14)$$

where "CH" denotes the Cauchy principal value of this integral.

In a semiconductor, the absorption is in some sense the opposite to luminescence: A photon passing through the material may be absorbed by transferring its energy to a carrier which is thus lifted to an excited state (c.f. Eq. 6.1).

Therefore, the density of states $D(E)$ can be probed by measuring the absorption coefficient $\alpha(\lambda)$. Thus, in a direct (bulk) semiconductor, the absorption near the band edge is described by

$$\alpha(\hbar\omega) \sim (\hbar\omega - E_{\mathrm{g}})^{1/2}$$

because $D(E) \sim (E - E_{\mathrm{g}})^{1/2}$. Similarly, absorption spectra reflect the qualitatively different density of states in lower dimensional systems (see Chapter 10).

In an indirect semiconductor, a third particle, typically a phonon, is required to fulfill both conservation of energy and momentum (see Section 1.9). Hence, the absorption near the band edge reads

$$\alpha(\hbar\omega) \sim (\hbar\omega \mp E_{\mathrm{Phonon}} - E_{\mathrm{g}})^2.$$

Consequently, two "gaps" may be detected in absorption measurements: $E_{\mathrm{g}} + E_{\mathrm{Phonon}}$ if a phonon is emitted or $E_{\mathrm{g}} - E_{\mathrm{Phonon}}$ if a phonon is absorbed. The latter case requires the presence of phonons, i.e., it is typically visible only at higher temperatures (e.g., room temperature). The lowest phonon energy in Si which fulfils these boundary conditions is 18 meV (TA phonon).

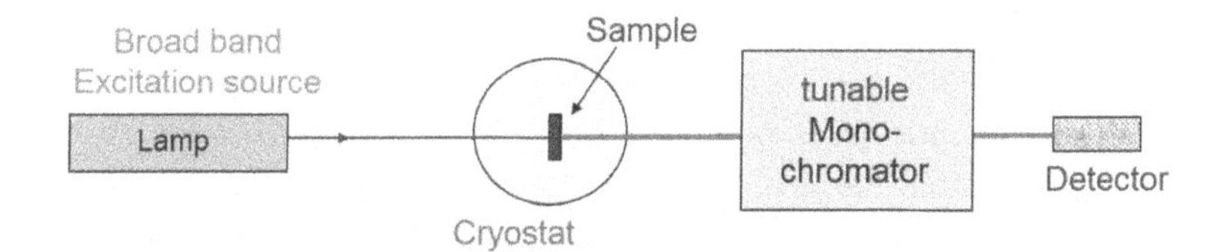

Figure 6.5 Absorption measurement setup (schematically). Notice that the monochromator is between sample and detector in order to separate any parasitically generated photoluminescence signal from the transmitted light.

Besides the density of states, other states may also be visible in absorption (e.g., impurities, excitons), if their density is large enough.

Absorption measurement setup

In principle, the same components as in PL can be used to measure absorption. They just have to be arranged differently (Fig. 6.5). The light source now must be a broad band source or tunable in wavelength. What is measured is not the absorption directly but the intensity of the transmitted light. Therefore, corrections are necessary due to the light lost by reflection.

Although it seems best to place the monochromator between light source and sample, this leads to the problem that not only the transmitted light enters the detector, but also all excited photoluminescence. This can be blocked by placing the monochromator between sample and detector.

This method works nicely for strong absorption. However, in the case of weak absorption (e.g., absorption of thin layers like quantum wells or impurity levels of not too high concentration), the transmitted light I_T (of fairly high intensity) is nearly the same as the incoming light I_i. Hence, the absorption signal

$$I_A \approx I_i - I_T \quad \text{(neglecting reflected intensity)}$$

and therefore the signal-to-noise ratio may be very small. Then, other methods may do a better job, e.g.,

- **Photoluminescence excitation spectroscopy:** Here, photoluminescence is measured on one fixed (representative) wavelength (e.g., main emission of a quantum well from recombination of the ground states in CB and VB), while the exciting light is tuned in wavelength. Now either a strong lamp or a tunable laser is required, because only a small wavelength band should shine on the sample. The respective signal is a convolution of absorption and the respective relaxation processes connecting the excited primary carriers (e.g., their excitation into higher quantum well sub-bands) with the PL process. In many cases, the relaxation is much faster than the final recombination process and thus the final signal is a good representation of the absorption signal. This method can also help in cases where a strong substrate absorption would inhibit any transmission measurement.
- **Photothermal deflection spectroscopy:** Here, one makes use of the fact that any absorption typically results in a heating of the sample. Even small heating can be measured as follows: The sample is immersed into a liquid the refractive index of which changes strongly with temperature. A probe laser beam is carried very close parallel to the sample surface (Fig. 6.6). If the sample and consequently the liquid on its surface heats up

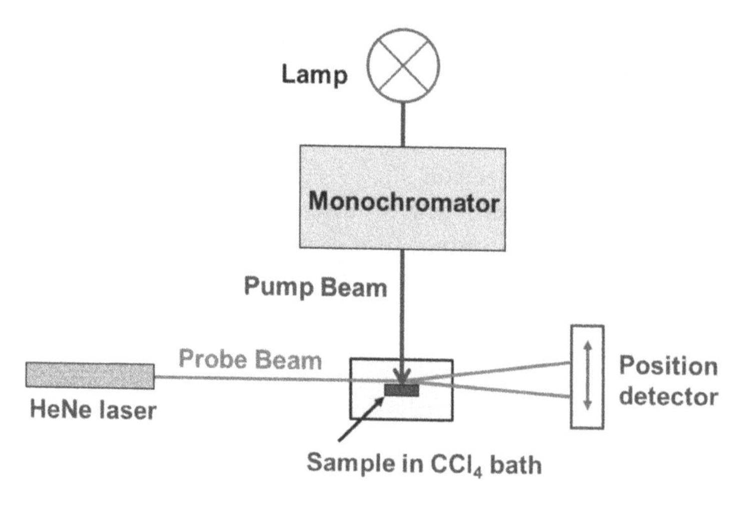

Figure 6.6 Photothermal deflection spectroscopy setup (schematically).

due to absorption, the laser beam is slightly deflected, which can be measured with a respective sensor.

6.3.3 Lateral Absorption and Gain Spectroscopy

In many cases, we are interested in the properties of light traveling in the epitaxial structure parallel to the sample surface, e.g., in waveguiding structures. This may be measured by exciting some luminescence in a point not too far from the sample edge and measure the light which exits at the wafer edge. By varying the distance between the excitation point and the edge and relating to the measured intensity, the respective absorption may be deduced.

Another key property of such waveguiding structures is their optical gain, which eventually is the key issue of any laser action. As mentioned above, the light gets amplified by stimulated emission: A photon triggers an excited carrier to recombine thus creating a twin photon. Both photons may trigger further processes and so on. This means that the light intensity increases exponentially with the traveling distance, which can be described in strong analogy to absorption (see Eq. 6.13):

$$I(x) = I(0)e^{gx}, \tag{6.15}$$

where g is the gain coefficient, which depends on the specific material and heterostructure details and also on the excitation strength (see Section 12.2.1).

This feature can be measured with a similar method: Here, the sample is excited in stripe geometry. The stripe is reaching up to the sample edge (see Fig. 6.7). For strong excitation, stimulated emission will occur. The intensity of the light exiting at the edge (amplified spontaneous emission, ASE) depends on the length of the stripe according to

$$I(L,\lambda) = \frac{\beta r_{\text{spont}}}{g}(e^{g\cdot L} - 1) = I(0)(e^{g\cdot L} - 1).$$

By varying the stripe length, the gain coefficient $g(\lambda)$ can be evaluated from this equation, which is an important parameter particularly for laser diode structures.

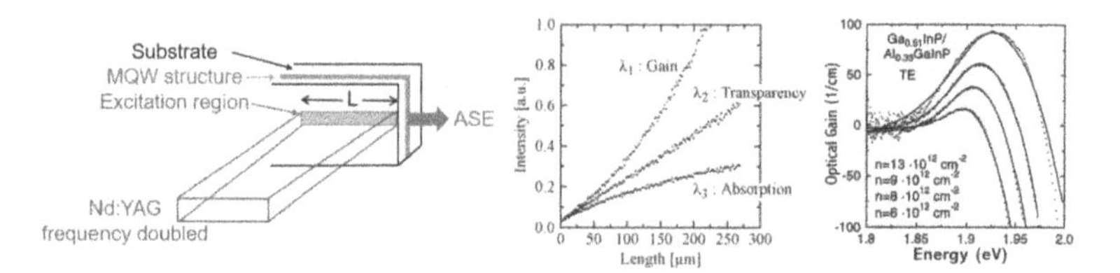

Figure 6.7 Optical gain spectroscopy setup schematically using the stripe length method (left) and typical results for different excitation conditions (middle). At a wavelength far from the gain maximum, absorption dominates, whereas at optimum wavelength, stimulated emission may dominate. (Right) gain spectra versus energy for a GaInP quantum well structure for various excitation intensities (after [46]).

6.3.4 Fourier Spectroscopy

In classical spectroscopy (as discussed above), the intensity of the light selected by the monochromator hitting the detector is very low, particularly if high spectral resolution is required which only can be achieved by narrow entrance and exit slits of the monochromator (see Section 6.4). This problem can be overcome by **Fourier spectroscopy**: The basic idea is to illuminate sample and detector by the full power of the light to be analyzed, i.e., its full (broad) wavelength range. Now, the sample signal and a reference signal are brought to interference in a Michelson interferometer (see Fig. 6.8). The signal intensity as a function of the position of the moveable mirror $I(x)$ is the Fourier transformation of the wanted signal $S(\lambda)$:

$$I(x) = \int_{-\infty}^{\infty} S(\lambda)e^{-\lambda x} d\lambda$$

Hence, the latter can be calculated from the former. Special calculation techniques have been developed for a fast and efficient Fourier transformation.

This method hence has the so-called "multiplex advantage," as the complete signal spectrum is measured all the time in the detector. This is in particular good in the infrared range, as in this range detectors are fairly noisy.

Depending on the mounting of the sample, (more or less) all kinds of spectroscopy can be performed including photo-

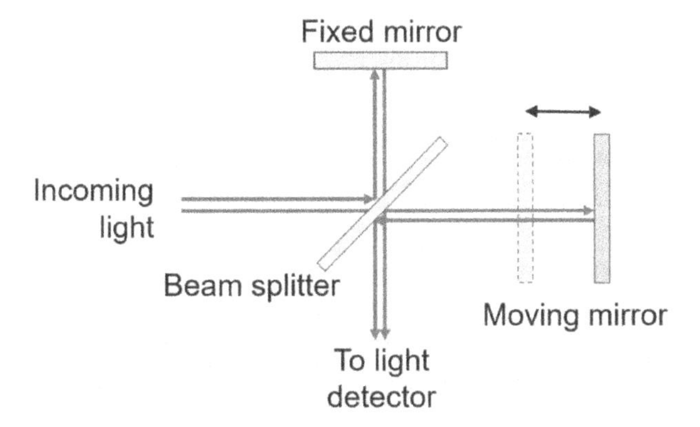

Figure 6.8 Michelson interferometer (schematically). The sample may be placed at different positions. When measuring absorption, the sample is placed in the incoming light beam (white spectrum). For photoluminescence, the sample itself "generates" the incoming light.

luminescence, absorption, electroluminescence, electroabsorption, photocurrent, etc.

Here, the resolution is determined by the maximum translation distance of the moving mirror, whereas the spectral range is given by the translation resolution.

6.3.5 Other Related Methods

There is a huge number of very specific spectroscopic methods which help to evaluate many more details of solid-state material properties. Let us just mention briefly some methods which are fairly closely related to the basic methods discussed above.

- Time resolved spectroscopy:
  In principle, any of these methods can be performed with a pulsed excitation and a time-resolved detection. This is in particular done in photoluminescence providing information about carrier lifetimes, etc. Depending on the light pulse and detection technique, time resolutions from sub-picoseconds upward can be achieved.
  Detection techniques to obtain good time resolution:
  - Single photon counting, time to amplitude conversion

 - Streak camera
 - Four-wave mixing, pump-probe.

- Modulated absorption, e.g., laser modulation, electro-absorption:
  This can be used to measure, e.g., absorption properties in structures out of thermal equilibrium, e.g., after laser excitation or after electrical excitation. Modulation in time may then be used to improve the signal-to-noise ratio.
- Luminescence microscopy:
  By detecting the local position of the luminescence in an optical microscope, specific local information of spectroscopic properties can be obtained.

6.4 Some Technical Details

- Monochromator: In order to determine the wavelength of the spectroscopic signal, typically, the diffraction at a reflection grid is taken as wavelength selective device, see Fig. 6.9. By rotating the diffraction grid, the wavelength in the beam which hits the exit slit fulfilling constructive interference changes.

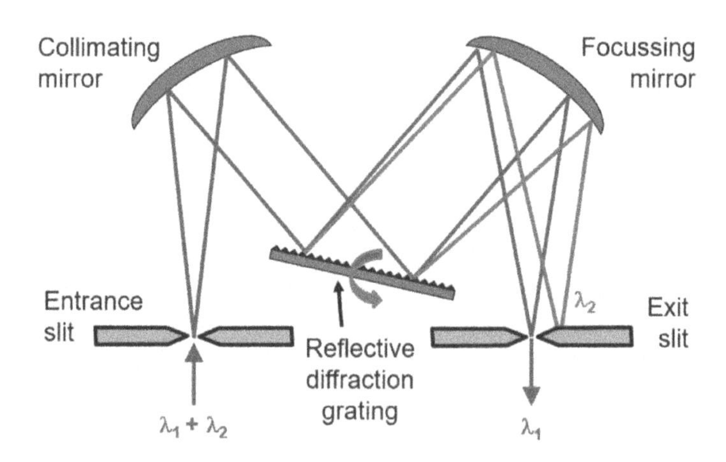

Figure 6.9 Folded monochromator: Czerney–Turner principle. By decreasing the slit opening, the spectral resolution can be increased. However, this decreases simultaneously the signal to noise ratio.

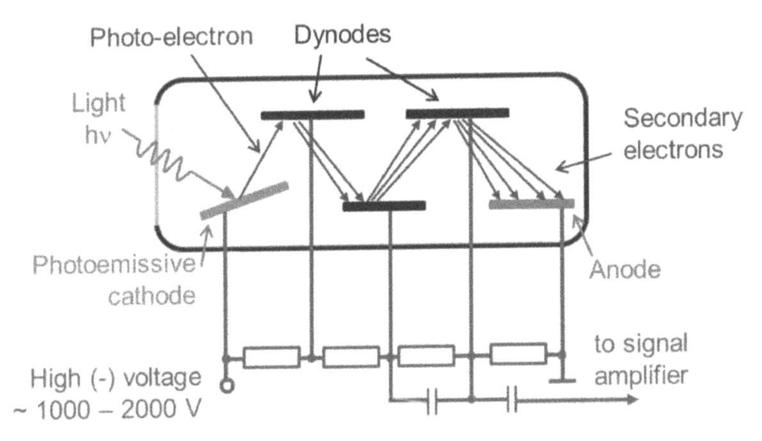

Figure 6.10 Photo multiplier. Typical data: Wavelength range: 110–1100 nm; quantum efficiency of photo cathode: 1–10 %, response time: 1–20 nsec

Hence, the spectral resolution is determined by the slit width and the length of the optical path.

- Detector: The main tasks of a light detector are a linear response in terms of intensity and constant response concerning wavelength, both over a wide range. Different device technologies are used, to some extent depending on the required wavelength range:
  - Photomultiplier (Fig. 6.10). The light arrives on the photoemission cathode, i.e., the photon generates an electron. This electron can be accelerated in an electric field before it hits another cathode. Due to its gained energy, it will release several new electrons which are accelerated to the next cathode. Thus, a strong amplification is achieved.
  - Semiconductor diode (pin diode), which in principle works like a solar cell (see Section 12.3). By the applied negative voltage, the carrier separation after absorption of photons can be enhanced. Depending on semiconductor material, different wavelength ranges can be addressed (see Fig. 6.11).
  - Charge coupled device (Fig. 6.12): Integrated-circuit chip consisting of an array of capacitors that store charge when light creates electron-hole pairs. The charge accumulates

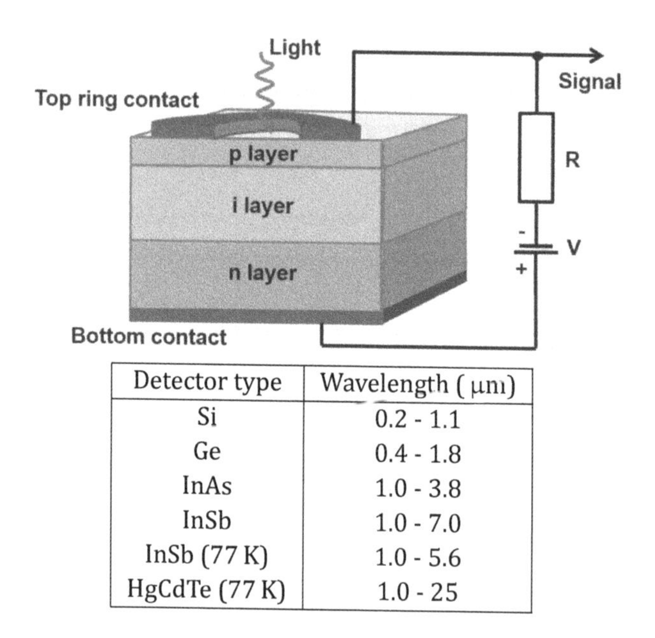

Detector type	Wavelength (μm)
Si	0.2 - 1.1
Ge	0.4 - 1.8
InAs	1.0 - 3.8
InSb	1.0 - 7.0
InSb (77 K)	1.0 - 5.6
HgCdTe (77 K)	1.0 - 25

Figure 6.11 Semiconductor diode detector (top) and wavelength range (bottom).

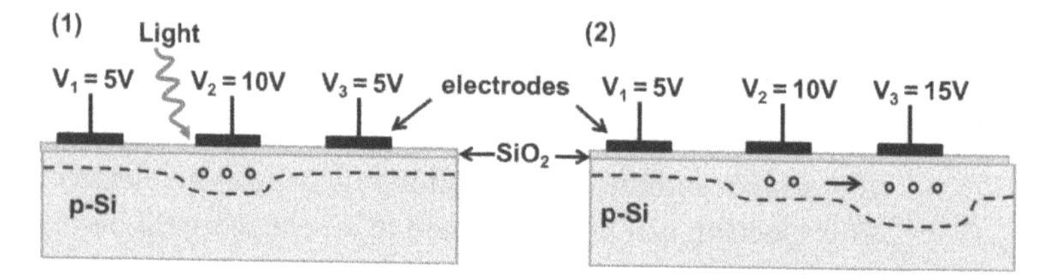

Figure 6.12 Three cells of a charge coupled device (schematically). Once carriers are created at a pixel, they can be shifted step by step by changing the voltage to the adjacent pixels adequately (compare left and right picture) to be eventually read out.

and is read in a fixed time interval. With multi-pixel devices, a simultaneous read-out of a complete spectrum can be done (e.g., 1024 pixels) making the acquisition of a spectrum very fast. It is highly sensitive.

- Lock-in amplifier: Typically, the signal-to-noise ratio of a spectroscopic signal is improved by a lock-in-amplifier. This device provides a phase-sensitive rectification of the signal. If the sample excitation is pulsed (e.g., by a chopper wheel), then the generated signal shows the same pulsing. By the phase-sensitive rectification, the signal is only positively amplified then when it is valid. Thus, any noise which does not coincide with the pulse frequency and phase (like steady-state background light) is suppressed.

Problems

(1) Basic optical processes:
   (a) What do you need as input for the three basic optical processes, and what is the output?
   (b) What is the difference between spontaneous and stimulated emission?
   (c) Correlate the three basic optical processes with the devices where these processes are applied.

(2) Explain: When can we speak about "quasi Fermi levels"? What are the requirements to make such description meaningful?
(3) Explain: What is meant by the terminus "ABC model" regarding recombination rates in semiconductors?
(4) Explain the "Auger effect."
(5) (a) What is an exciton?
   (b) Why is it typically not relevant at room temperature?
(6) (a) Which kinds of luminescence may be obtained in compound semiconductors? Explain the differences between these methods.
   (b) Explain the setup of a photoluminescence experiment.
   (c) What can be measured by luminescence at low temperature?
(7) How can you measure the band gap of a semiconductor?
(8) Describe qualitatively a photoluminescence and absorption spectrum.
(9) What is the importance of the "Kramers-Kronig relation"?
(10) Where should the monochromator be placed in an absorption experiment? Why?

(11) (a) Why is it difficult to obtain directly clear absorption data of a very thin layer?
(b) What method would help to solve this problem? Explain how it works. What are the differences compared with a classical absorption experiment?

(12) Explain how gain spectroscopy works.

(13) What is the advantage of a Fourier spectrometer?

Chapter 7

X-Ray Diffraction

Owing to the very small lattice constants of inorganic crystals (fractions of nanometers for our compound semiconductors), their crystalline lattice properties cannot be analyzed with optical microscopy, but illumination with much shorter wavelength is required. Therefore, an important tool to characterize compound semiconductors and respective epitaxial structures is **high-resolution x-ray diffraction** sometimes abbreviated as **HRXRD**, with which very important information can be obtained like

- lattice constants,
- chemical composition of ternaries, etc.,
- strain,
- quality, dislocation density, etc., and
- superlattice period,

as briefly explained in the following sections.

7.1 Basics

The principle of x-ray diffraction is explained in Fig. 7.1:

A monochromatic x-ray beam with a wavelength of a few 100 pm is reflected at crystal planes according to Bragg's law: The beam

Compound Semiconductors: Physics, Technology, and Device Concepts
Ferdinand Scholz

ISBN 978-981-4774-07-9 (Hardcover), 978-1-315-22931-7 (eBook)
www.panstanford.com

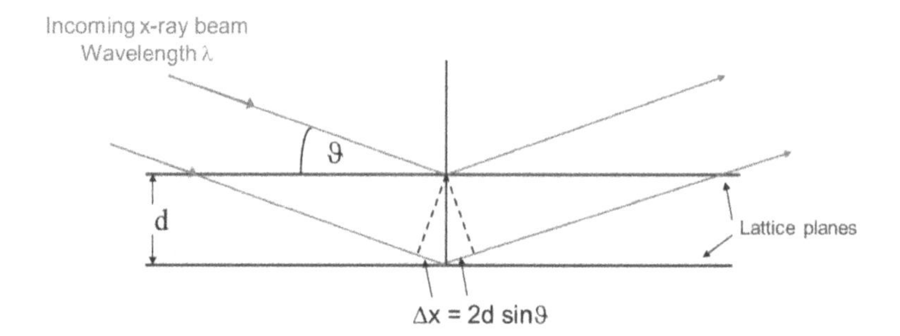

Figure 7.1 Bragg's law of diffraction.

reflected at the lower interface must travel a longer distance than the upper beam. Therefore, constructive interference of the reflected x-ray beams at parallel lattice planes having a spacing d and an angle of incidence of ϑ occurs if

$$n \cdot \lambda = 2d \sin \vartheta, \tag{7.1}$$

where n describes the order of constructive interference.

The distance of parallel crystal planes can be calculated from their respective Miller's indices according to

$$d = \frac{a}{\sqrt{h^2 + k^2 + l^2}} \quad \text{for cubic material and}$$

$$d = \left(\frac{4}{3} \frac{h^2 + k^2 + hk}{a^2} + \frac{l^2}{c^2} \right)^{-1/2} \quad \text{for hexagonal material.}$$

As discussed earlier (see Section 1.3.2), this can be also described by Laue's law of diffraction considering the atomistic nature of the crystals. Then we get **Laue's equations**:

$$\vec{k_1} - \vec{k_0} = n\vec{G}, \tag{7.2}$$

where $\vec{k_0}, \vec{k_1}$ are the impinging and the exiting x-ray beam, respectively with $|\vec{k_0}| = |\vec{k_1}|$ to fulfill energy conservation (elastic reflection), and $\vec{G}$ is a vector of the reciprocal lattice.

Equation 7.2 describes a sphere in the reciprocal space with radius $|\vec{k_0}|$ (**Ewald's sphere**). We see: Constructive interference only occurs for those directions where the circumference of Ewald's sphere hits a reciprocal lattice point.

The **intensity** of the x-ray beam is described by the structure factor F, which depends on the detailed position r_j of the atoms in the crystalline unit cell and of the atomic properties:

$$F = \sum_{j=1}^{s} e^{-i\vec{G}\vec{r}_j} \cdot f_j, \tag{7.3}$$

with s = number of atoms in unit cell (zinc blende: $s = 8$, i.e., four atoms of group III and four atoms of group V) and f_j = atomic form factor.

Constructive interference is obtained for

$$F = f_{III}[1 + e^{-i\pi(h+k)} + e^{-i\pi(k+l)} + e^{-i\pi(l+h)}] + f_V[\ldots] \neq 0. \tag{7.4}$$

Depending on the involved crystal planes, hence the order of diffraction, we obtain:

$$\begin{aligned} F &= 0 \quad \text{for} \quad 100, 300, 500, \ldots \\ &= 4(f_{III} - f_V) \quad \text{for} \quad 200, 600, \ldots \rightarrow \text{small} \\ &= 4(f_{III} + f_V) \quad \text{for} \quad 400, \ldots \rightarrow \text{large} \\ \text{or} &\neq 0 \quad \text{for} \quad h, k, l \text{ even}, h + k + l = 4n \\ &\neq 0 \quad \text{for} \quad h, k, l \text{ even}, h + k + l = 4n + 2 \\ &\neq 0 \quad \text{for} \quad h, k, l \text{ uneven} \\ \text{otherwise} &= 0. \end{aligned}$$

Please notice: These results just reflect the choice of a cube as unit cell for the zinc blende crystals. (400) reflection then just means that the x-ray beam is reflected constructively at *every* crystal plane (being group III or group V). There are 4 such planes in one cube unit! Therefore, this order of diffraction is typically the most intense.

7.2 What Can Be Measured?

- **Lattice constant**
  The lattice constant is directly related to the composition of ternary and multinary layers via Vegard's law (Eq. 2.1). However, to get an accurate measurement of the chemical composition, the lattice constant has to be measured with high accuracy.

Example: The lattice constants of GaAs ($a = 0.565$ nm) and InAs ($a = 0.606$ nm) differ by about 7%. If we like to determine the composition x of $Ga_{1-x}In_xAs$ within an accuracy of 1%, then the lattice constant must be measured better than 7×10^{-4}. This can be transformed to the accuracy of the respective diffraction angle difference $\Delta\vartheta$ (for GaAs (004)) of about ± 90 arcsec. This requires an excellent goniometer, etc., and an extremely precise mounting of the sample.
Particularly the latter requirement is less important, if the layer to be analyzed is epitaxially grown on a layer/substrate with well-known lattice constant, e.g., some binary material. Then, we can evaluate easily the difference of the two lattice constants from the difference of the two Bragg angles. For the normal case of small lattice mismatch, the Bragg angles of the substrate and of the epitaxial layer to be examined are nearly the same ($\vartheta_{\text{substrate}} \approx \vartheta_{\text{layer}} \approx \vartheta$). Then we can obtain the lattice mismatch as

$$\frac{\Delta a}{a} = -2 \cot\vartheta \sin\frac{\Delta\vartheta}{2}$$

$$\text{with} \quad \Delta\vartheta = \vartheta_{\text{layer}} - \vartheta_{\text{substrate}}.$$

If $\Delta\vartheta$ is very small, then we can further simplify:

$$\frac{\Delta a}{a} = -\Delta\vartheta \cot\vartheta. \tag{7.5}$$

Now, just the relative angle must be measured with high accuracy (typically at least $\Delta\vartheta < 20$ arcsec) (see Fig. 7.2).

- **Strain** In many cases, the crystal lattice of our epitaxial layers may be deformed by external forces leading to a strained situation. As discussed in Chapter 9, typical reasons for strain are as follows:
  - **Lattice mismatch**: If a layer of only slightly different lattice constant is grown on a substrate, then it may grow so that the layer takes over the in-plane lattice constant $a_{||}$ of the substrate (**pseudomorphic growth**). Hence, it gets biaxially strained which results in an opposite change of the perpendicular lattice constant $a_{\perp}$. This works up to some critical thickness depending on the lattice mismatch, governed by the balance of forces or energies (strain versus creation of dislocations). Typical values: Some 10 nm for a strain $\frac{\Delta a}{a} \sim 1\%$.

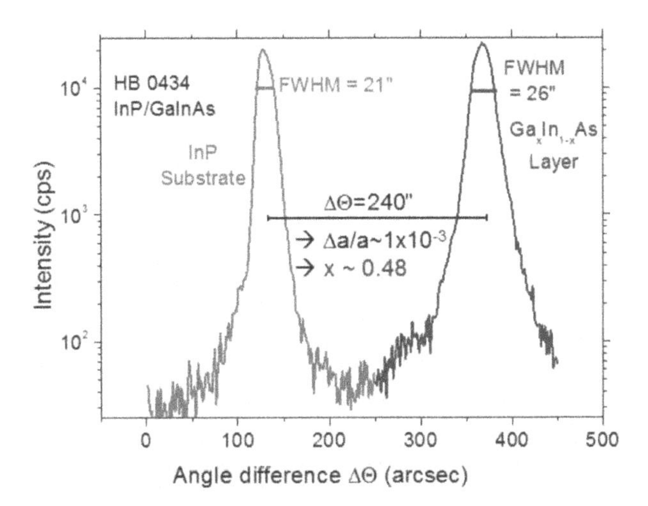

Figure 7.2 HRXRD signal of a $Ga_xIn_{1-x}As$ layer grown on InP. From the distance of the two peaks, the lattice mismatch and hence the composition can be obtained.

- **Thermal mismatch**: If a layer grows unstrained at growth temperature (several 100°C) on a substrate, it may get strained during cooling down due to the different thermal expansion coefficient α as compared to the substrate. This also leads to biaxial strain changing the in-plane and the out-of-plane lattice constants in opposite directions. Typical values: $\frac{\Delta a}{a} \sim 0.1\%$.

To analyze this situation, both (or even all three) lattice constants need to be determined.

In standard x-ray measurement, the lattice planes to be evaluated are parallel to the sample surface. Then the incident angle ω and the exiting angle ϑ are equal (see Fig. 7.1). Hence, only the lattice constant perpendicular to the surface can be determined.

However, the in-plane lattice constant of an epitaxial layer cannot be measured so easily by HRXRD. Therefore, so-called asymmetric reflexes need to be measured, which contain signals from both lattice constants. A typical example is the (115) reflection of GaAs (Fig. 7.3).

Notice: The lattice planes of substrate and epitaxial layer are no longer parallel for the asymmetric configuration. To

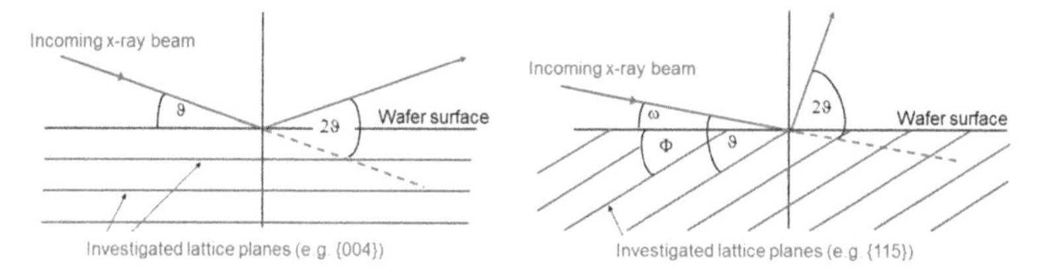

Figure 7.3 Symmetric (left) and asymmetric (right) x-ray diffraction measurement.

obtain the two unknowns $a_{||}$ and $a_{\perp}$, two measurements are necessary, e.g., both directions (exchange incoming and outgoing beam).

Moreover, an exact wafer orientation (rotation with respect to the axis perpendicular to the surface) is necessary to adjust the plane spanned by the incoming and outgoing beam perpendicular to the respective lattice planes.

In most cases, the asymmetric beams have lower intensity, as the respective lattice planes are occupied by less atoms.

- **Composition:**

  As mentioned above, the composition is related to the lattice constant via Vegard's law (Eq. 2.1). Hence, it can be evaluated in particular for ternary compounds. For quaternary compounds and higher, more measurements are needed (e.g., band gap by optical spectroscopy). Therefore we focus on ternaries in the following explanation.

  In many cases, the epitaxial layers are strained as discussed in the last paragraph. This must be taken into account when applying Vegard's law.

  If the layers are pseudomorphically strained ($(\frac{\Delta a}{a})_{||} = 0$), then the lattice constant $a_{\perp}$ measured in the symmetric arrangement is related to the lattice constant a_{relaxed} of the unstrained material by

$$\left(\frac{\Delta a}{a}\right)_{\perp} = \left(\frac{\Delta a}{a}\right)_{\text{relaxed}} \cdot \frac{1+\nu}{1-\nu},$$

  where ν = Poisson's number, i.e., a characteristic elastic constant of the material.

For cubic phosphides and arsenides: $\nu \simeq 0.3$; hence,

$$\left(\frac{\Delta a}{a}\right)_{\perp} \simeq 2 \cdot \left(\frac{\Delta a}{a}\right)_{\text{relaxed}}.$$

If the material is not completely pseudomorphically strained, then $a_{\perp}$ **and** $a_{||}$ must be measured (as described above) to get an accurate value for the relaxed lattice constant.
Hence, the strain situation must be carefully analyzed before extracting a composition value from XRD results.

- **Superlattice period:**
  A periodic sequence of thin epitaxial layers is called a superlattice. As the period of such a superlattice (typically some 10 nm) is still not far from the x-ray wavelength, it can also be measured by HRXRD by analyzing the constructive interference at every superlattice period. Typically, a superlattice is composed of a thin layer of thickness L_z of a material with lower band gap followed by another thin layer with thickness L_B of a material with larger band gap forming a period $P = L_z + L_B$; this sequence is repeated several times (see also Chapter 8).
  The period can be put into Braggs's law (Eq. 7.1) to evaluate the angles of constructive interference. As the period is quite large compared to the x-ray wavelength, many modes can be detected, the so-called *superlattice peaks* (Fig. 7.4). Hence, the period of the superlattice can be evaluated from the distance of two superlattice peaks $\Delta\vartheta$ (i.e., n_{th} and $(n+1)_{th}$ order) for the symmetric case:

$$P = \frac{\lambda}{\Delta\vartheta} \frac{1}{2\cos\vartheta},$$

  where ϑ = the respective incoming angle.
  From the distance between the 0_{th} satellite peak and the substrate reference peak, the material and hence the composition of the quantum well can be deduced (if the barrier is the same material as the reference) by taking into account the lower amount of quantum well material in the superlattice:

$$\left(\frac{\Delta a}{a}\right)_z^{QW} = \left(\frac{\Delta a}{a}\right)_z^{\text{measured}} \cdot \frac{L_z + L_B}{L_z}$$

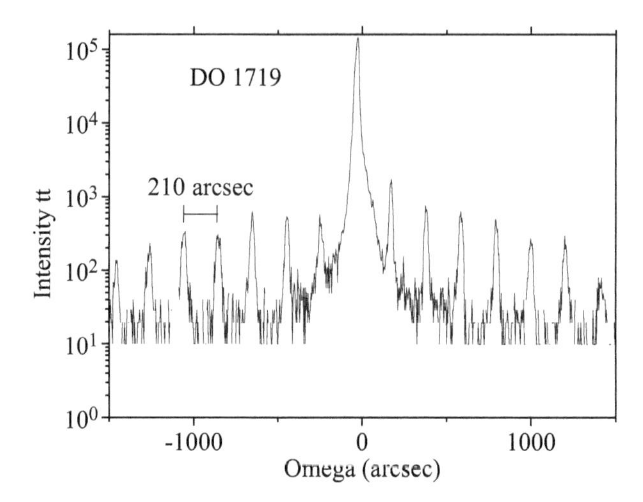

Figure 7.4 HRXRD signal of a GaInAs-InP superlattice. From the distance of the satellite peaks ($\Delta\vartheta = 210$ arcsec) a superlattice period of $P = 90$ nm can be deduced.

Similarly, in simple structures, the layer thickness can be determined by the interference of the reflections at the upper and the lower interface:

$$d = \frac{\lambda}{\Delta\vartheta}\frac{\vartheta + \Phi}{\cos 2\vartheta}$$

Here, $\Phi =$ inclination of respective lattice planes to the sample surface.

However, these interferences are only visible for perfect interfaces. They are called **Pendellösung fringes**.

- **Layer quality:**
  The measured x-ray diffraction peak **line width** is an indication for the **layer quality**. It is closely related with the crystalline defect density, the alloy composition fluctuations, etc. If line defects like threading dislocations are the main reason for a x-ray diffraction line broadening, then the dislocation density N_D may be evaluated from the linewidth $\Delta\vartheta$ according to

  $$N_D = \frac{(\Delta\vartheta)^2}{9b^2},$$

  where b is the Burger's vector of the respective dislocation. Depending on the direction of the Burger's vector, different

Table 7.1 Effect of substrate and epilayer parameters upon the rocking curves (after [47])

Material parameter	Effect on rocking curve	Distinguishing features
Mismatch	Splitting of layer and substrate peak	Invariant with sample rotation
Misorientation	Splitting of layer and substrate peak	Changes sign with sample rotation
Dislocation content	Broadens peak	Broadening invariant with beam size No shift of peak with beam position on the sample
Mosaic Spread	Broadens peak	Broadening may increase with beam size up to mosaic cell size No shift of peak with beam position on the sample
Curvature	Broadens peak	Broadening increases linearly with beam size Peak shifts systematically with beam position on sample
Relaxation	Changes the splitting	Different effect on symmetrical and asymmetrical reflection
Thickness	Affects intensity of peaks. Introduces interference fringes	Integrated intensity increases with thickness up to a limit. Fringe period controlled by thickness
Inhomogeneity	Effects vary with position on sample	Individual characteristics may be mapped

types of dislocations may be analyzed by evaluating different asymmetric x-ray peaks.

Table 7.1 gives an overview over different results which can be obtained by high-resolution x-ray diffraction.

7.3 Diffractometer Equipment

The main components of a high-resolution x-ray diffraction system as required for the above explained measurements are briefly discussed in the following:

- **X-ray tube**: This is a vacuum tube, where electrons are accelerated to typically 20–60 keV before hitting the anode

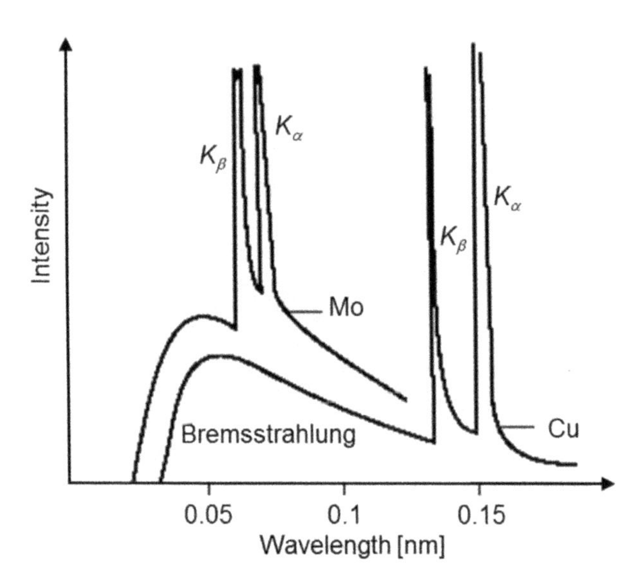

Figure 7.5 Typical emission spectrum of a copper (Cu) and a molybdenum (Mo) x-ray anode (from http://www.spektrum.de/lexikon/geowissenschaften/roentgenroehre/13744 (visited 2016, with permission of Springer). The copper anode was driven with a voltage of 38 kV corresponding to the lowest wavelength of the Bremsstrahlung at 0.0326 nm.

made of some specific material. In many cases, a copper (Cu) anode is used.

The high-energy electrons are thus interacting with the anode material. A broad spectrum of electromagnetic waves is generated by their deceleration, the so-called **Bremsstrahlung** which contains photons of energies up to the electron acceleration energy (Fig. 7.5). This is superimposed by some material specific lines.

The strongest line for Cu is the $Cu_{K\alpha}$-line: By the electron beam, an electron is removed out of the K-shell of a Cu atom. Subsequently, it is replaced by an electron of the L-shell. The energy difference of this process (about 8028 eV) results in the generation of a photon ($\lambda = 0.154$ nm). There are two major sub-levels in the L-shell; therefore the K_α line consists of two components: $K_{\alpha 1}$ and $K_{\alpha 2}$.

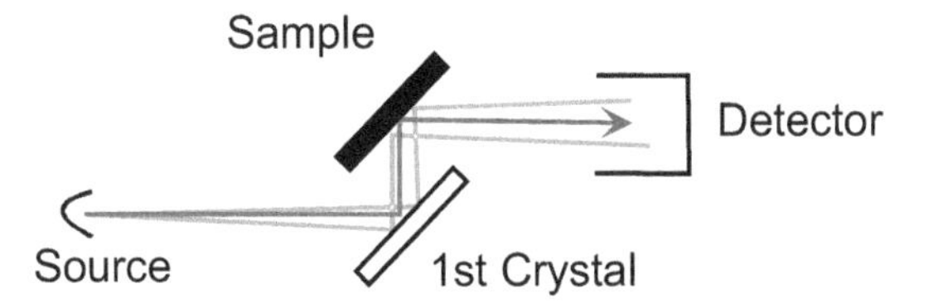

Figure 7.6 Double-crystal diffractometer, where the x-ray beam is reflected at one monochromator crystal and at the sample.

- A **monochromator** is necessary to select one specific wavelength, i.e., to remove all other wavelengths from the x-ray beam. A simple monochromator is just the same crystal as the crystal to be analyzed (→ double-crystal x-ray diffraction, Fig. 7.6). At a given angle, only that wavelength will be reflected which fulfills Bragg's law. Notice: Under slightly different angle, some other wavelengths are reflected which may also hit the sample. As they hit the sample under the same "wrong" angle, they contribute also to the final constructive interference as long as the first and the second crystal are the same.
  This increases the total intensity with still good angular resolution, as both, $K_{\alpha 1}$ and $K_{\alpha 2}$ contribute to the final signal. However, if the sample differs from the monochromator crystal, then the two Cu lines are both visible at different angles, which makes the obtained curves questionable. Hence, the flexibility of such an equipment to measure different materials is very limited.
  Therefore, a better solution was invented: A so-called "channel-cut crystal" acts as a monochromator (Fig. 7.7). The primary beam is reflected four times under constructive interference conditions (if angles are adjusted correctly). Hence, only one narrow wavelength peak survives. The four reflecting crystals need not to be adjusted separately with extremely high precision. By cutting a channel into a single crystal (resulting in a U-shaped crystal), two of the reflecting planes are already perfectly adjusted to each other, reducing the alignment problems significantly. The channel-cut crystal usually is made of Ge because of its high crystalline quality.

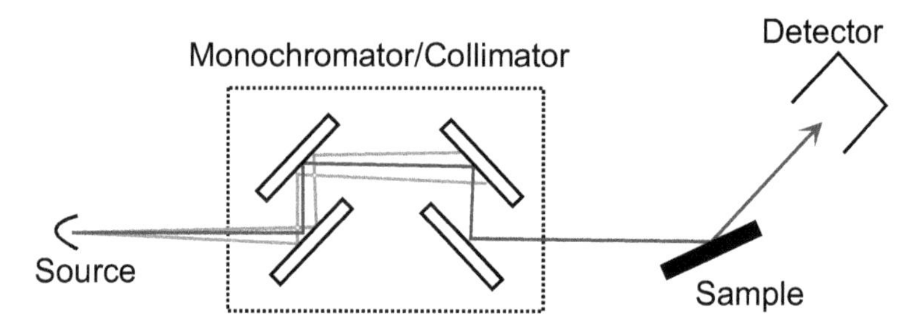

Figure 7.7 5 crystal diffractometer, where the x-ray beam is monochromized in four reflections in a channel-cut crystal before it goes to the sample.

- **Goniometers** are used to move the sample (or the x-ray source) and the detector precisely, hence controlling the incoming angle ω and outgoing angle 2ϑ. The movements are typically done with stepping motors with a resolution of few arcsec and high precision. Other axes like rotation and tilt of the sample typically can also be adjusted or even scanned. The same is true for the position of the sample (x-y-stepper) for wafer mapping.
- The x-ray beam is finally absorbed in the **detector**. It works similar to a photon detector (see Section 6.4). The area of the detector may be well defined by a slit (several 10–1000 μm) to increase the accuracy of the detection angle 2ϑ. Alternatively, an analyzer crystal is mounted between sample and detector. Depending on the detector arrangement, different information can be obtained:
    - Open detector: All x-ray intensity diffracted by the sample falls into the detector. Thus, any imperfection of the sample (variations in lattice constant as well as in planarity of the lattice planes) will broaden the signal (intensity versus ω).
    - The second most used mode is to move the sample by the angle ω and the detector by twice the angle $2\omega = 2\vartheta$. If an analyzer is installed, then the signal is broadened only by distance fluctuations of the lattice planes.
    - With an analyzer installed and fixed detector, the signal (intensity versus position ω of the sample with respect to

the incoming beam) is broadened by variations of the local inclinations of the lattice planes.

  - A two-dimensional scan (ω and ϑ independently), a so-called reciprocal space mapping provides information about strain and strain variations, etc. However, such a scan is fairly time-consuming. Find some more details in [48].

- Typically, a computer controls all movements and stores, evaluates, and displays the data.
- A shielding of the equipment is essential, as x-rays are harmful to human health.

Problems

(1) (a) Deduce Bragg's law.
    (b) Show for a one-dimensional point lattice that Laue's equation is equivalent to Bragg's law.
    (c) Explain: Why is the 004 peak of GaAs the most intense diffraction peak? Why does the 001 peak not exist?
(2) (a) Deduce the relation between lattice mismatch and difference of the Bragg angles of layer and substrate.
    (b) Do the same for the formula of the superlattice period.
(3) How can you determine the strain of a ternary semiconductor layer by x-ray diffraction? Explain.
(4) Explain how "reciprocal space mapping" is done. Which information can be deduced from such a measurement?
(5) What is the advantage of a channel-cut crystal in an XRD machine?

Chapter 8

Quantum Wells

Nowadays, very thin layers in a heterostructure bring essential advantages to many device functions, as discussed in this chapter. Their particular properties can be only understood by quantum physics. This is why they are called "quantum wells." You may find more details in many textbooks, e.g., in [4].

8.1 Semiconductor Heterojunction

Two different semiconductor materials in close contact, e.g., grown one on top of the other by some epitaxial method, form a **heterojunction**. Here, we focus first on junctions where the two materials have the same crystal structure and lattice constant, but differ in most other properties, in particular in the band gap E_g. Therefore, the crystal potential changes abruptly at the heterojunction requiring sophisticated calculations.

In a simplified picture, we assume the continuity of the vacuum level at the interface (c.f. Chapter 2, Fig. 2.4). Hence, the band offsets ΔE_c and ΔE_v can be evaluated.[a]

[a] In many cases, this simple rule does not give the correct values. Very often, the offsets have to be determined experimentally.

Compound Semiconductors: Physics, Technology, and Device Concepts
Ferdinand Scholz

ISBN 978-981-4774-07-9 (Hardcover), 978-1-315-22931-7 (eBook)
www.panstanford.com

Whatsoever, please notice: This change is very abrupt (less than a monolayer). Two to three monolayers already form a well-defined quantum well!

Depending on the size of the band offsets, different types of hetero-interfaces can be found (Fig. 2.5). We concentrate our further discussion on type I interfaces.

8.2 Quantum Well: Eigenstates

If a material of lower band gap is sandwiched between two layers of a material with larger band gap (forming for simplicity type I interfaces), then we call this structure a **quantum well structure**, if the thickness L_z of the center layer is fairly small (some ten nanometers, i.e., in the size of the de Broglie wavelength of the electron), because now the allowed energy states in this well must fulfill the Schrödinger equation with the boundary condition of the barriers on either side. This means that the respective wave function must fulfill some resonance conditions, similar to electromagnetic waves in a wave guide.

Let us determine the eigenstates of such a quantum well:

Such states are typically determined by applying the envelope wave function method. This assumes a potential $V(\vec{r})$ which only slowly varies with $\vec{r}$ compared to the crystal potential (caused by the regularly arranged atoms). Then it is a good approximation to put all effects of the crystal potential into the effective mass m^* of the carriers. It can be shown that also for the envelope function $\xi(\vec{r})$ the Schrödinger equation holds:

$$\left(\frac{p^2}{2m^*} + V(\vec{r})\right)\xi(\vec{r}) = (E - \varepsilon_0)\xi(\vec{r}) \tag{8.1}$$

with $\varepsilon_0 =$ the band edge of the respective band.[b]

Of course, at the heterojunction, the potential is certainly not slowly changing. However, it is possible to use this method for obtaining solutions for each layer, which then must be joined at the interfaces taking into account the correct boundary conditions.

---

[b]This simplification is only valid near $\vec{k} = 0$.

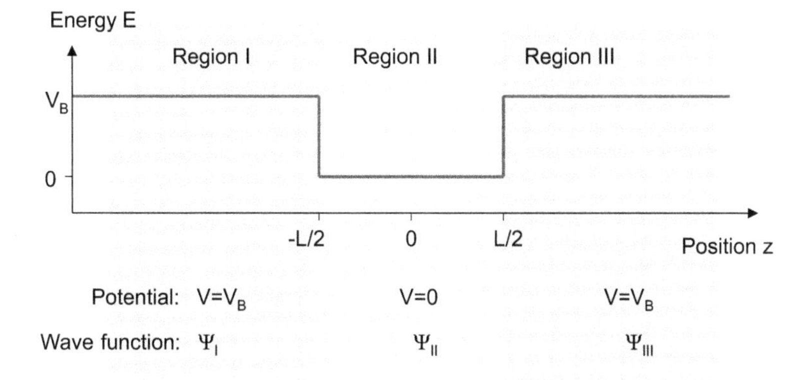

Figure 8.1 Energy scheme of a simple rectangular potential well.

Let us calculate a simple quantum well structure, i.e., a rectangular potential well as it may be formed for the electrons in the conduction band in a double heterostructure (Fig. 8.1), along these lines:

The Schrödinger equation now reads

$$\left(\frac{-\hbar^2}{2m^*}\Delta + V(z)\right)\Psi(\vec{r}) = E\Psi(\vec{r}), \tag{8.2}$$

$$\text{with} \quad \Delta = \frac{\partial^2}{\partial x^2} + \frac{\partial^2}{\partial y^2} + \frac{\partial^2}{\partial z^2}$$

$$\text{and} \quad V(z) = \begin{cases} 0 & \text{in the quantum well, i.e., for } -L_z/2 < z < L_z/2; \\ V_B & \text{in the barrier.} \end{cases}$$

In the quantum well, we take $m^* = m_\mathrm{W}$ (effective mass of the well material), whereas in the barrier material, the carriers typically have a different mass $m^* = m_\mathrm{B}$.

Please notice: $V(z)$ does not depend on x and y. Therefore, a separation is possible:

$$\Psi(\vec{r}) = \varphi(z) \cdot e^{i\vec{k}\vec{R}}, \tag{8.3}$$

with the two-dimensional vectors $\vec{R} = (x, y)$ and $\vec{k} = (k_\mathrm{x}, k_\mathrm{y})$.

The solution in the (x, y)-plane is the same as for a free particle:

$$E_{x,y} = \frac{\hbar^2 k^2}{2m^*}. \tag{8.4}$$

The following equation remains to be solved:

$$\left(\frac{-\hbar^2}{2m^*}\frac{d^2}{dz^2} + V(z)\right)\varphi_n(z) = \varepsilon\varphi_n(z) \tag{8.5}$$

This must be done in all three regions. As usual, we choose the ansatz

$$\varphi(z) = \alpha e^{\beta z},$$

which leads us to the characteristic equation

$$\frac{-\hbar^2}{2m^*}\beta^2 + V = \varepsilon. \tag{8.6}$$

In regions I and III, where $m^* = m_B$, we get

$$\beta = \frac{1}{\hbar}\sqrt{2m_B(V_B - \varepsilon)} = \lambda. \tag{8.7}$$

This is a real number (as long as $\varepsilon < V_B$) hence resulting in a damped function.

In region II, where $m^* = m_w$, we obtain

$$\beta = \frac{i}{\hbar}\sqrt{2m_w\varepsilon} = i\omega, \tag{8.8}$$

hence an imaginary number describing an oscillatory solution of a free-moving particle.

The following boundary conditions must be fulfilled:

(1) Normalization must be possible, i.e.,

$$\int_{\infty}^{\infty} \varphi^2(z)dz = 1.$$

This can only be fulfilled for $\varphi(\pm\infty) = 0$ in regions I and III, i.e., the solution is a damped exponential function in either barrier.

(2) The solutions must match at the hetero-interface ($z = \pm L_z/2$):

- The wave function must be continuous, i.e., $\varphi_B(\pm L_z/2) = \varphi_w(\pm L_z/2)$
- Moreover, the probability flux crossing the interface must be conserved, i.e.,

$$\frac{1}{m_B}\frac{d}{dz}\varphi_B(\pm L_z/2) = \frac{1}{m_w}\frac{d}{dz}\varphi_w(\pm L_z/2)$$

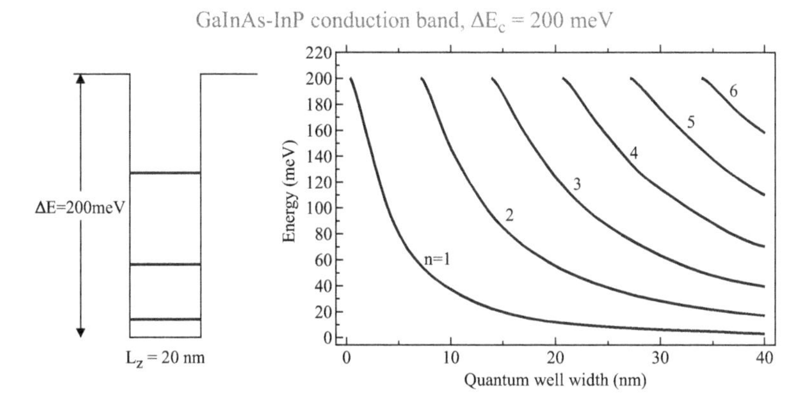

Figure 8.2 Solutions of Eqs. 8.9 and 8.10 for a GaInAs-InP quantum well for the electrons in the conduction band.

This leads to the following two characteristic equations for the energy eigenvalues ε:

$$\omega \tan\left(\omega \frac{L_z}{2}\right) = \frac{m_W}{m_B}\lambda \tag{8.9}$$

$$\omega \cot\left(\omega \frac{L_z}{2}\right) = -\frac{m_W}{m_B}\lambda. \tag{8.10}$$

This system of transcendent equations can be solved graphically or numerically. An example result is plotted in Fig. 8.2.

In many cases, a fair approximation may be obtained by taking $V_B = \infty$ (infinite square well potential). Then, the problem can be solved analytically leading to

$$\varepsilon_n = \frac{\hbar^2}{2m^*}\left(\frac{n\pi}{L_z}\right)^2. \tag{8.11}$$

Notice:

- The allowed particle energy is a function of the quantum well width L_z.
- There are several (n) allowed energy states, so-called quantum well **sub-bands**.
- For finite barrier height, the number n of sub-bands decreases with decreasing L_z and increasing ground state energy ε_1. However, there is always one solution ε_1, i.e., one bound state (in a symmetric quantum well)!

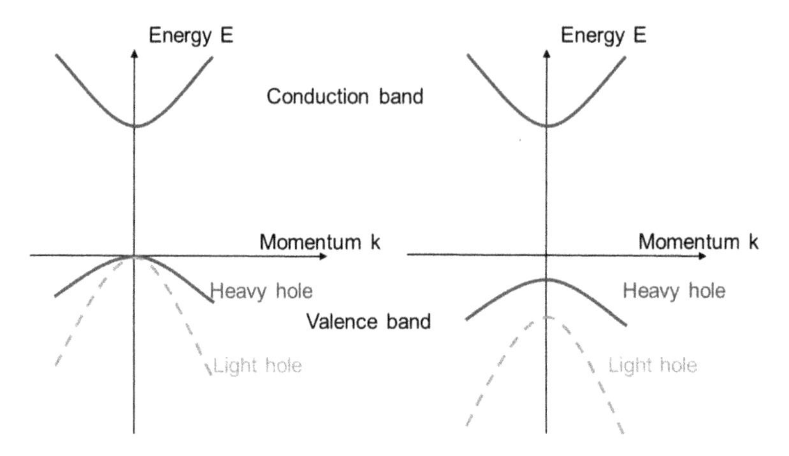

Figure 8.3 Lifting of the valence band degeneracy by different quantization of the heavy hole and light hole bands (right) compared to the bulk situation (left).

Exactly the same calculations can be performed for such potential wells in the valence band by taking into account the respective hole properties, in particular their effective mass.

As already discussed: Typically, the valence band is more complicated than the conduction band. For cubic III-V compound semiconductors it consists of a **heavy-hole**, a **light hole**, and a **split-off** band (c.f. Section 1.7). According to the different masses, we get two sets of solutions for the heavy and light hole, respectively, although they are degenerated in the bulk at $\vec{k} = 0$. In other words: **The degeneracy is lifted by the quantization** (see Fig. 8.3).

This can be also understood in an even more general picture: Typically, such a degeneracy is lifted by the break of some symmetry. Compare magnetic quantum numbers in an atom: The degeneracy of the respective orbits is lifted by an external magnetic field which acts as the symmetry breaker.

Here, the three-dimensional symmetry is broken by the definition of the quantum well plane.

Now, different new energetic levels are present in a quantum well. As the dispersion relation in the quantum well plane ((x, y)-plane) is, as for the bulk, still a parabolic relation (see Eq. 8.4), every

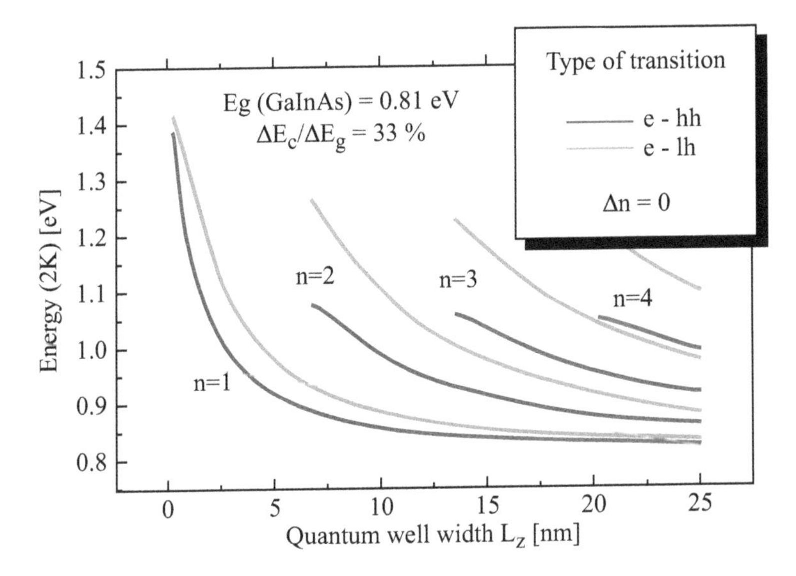

Figure 8.4 Transition energies in a GaInAs-InP quantum well at $T = 2$ K.

sub-band consists of such a two-dimensional parabola (see Fig. 8.5, middle).

For optical transitions, let us consider here (as for bulk material) only transitions at $\vec{k} = 0$, i.e., from the minima in the conduction sub-bands to the maxima in the valence sub-bands.

Carriers now can only occupy these sub-bands. Optical transitions are described by the energy difference between them:

$$\Delta E = E_{\mathrm{g}} + \varepsilon_{\mathrm{n}}^{c} + \varepsilon_{n'}^{v}, \tag{8.12}$$

i.e., even for the lowest sub-bands, the transition energy is larger than E_{g}. This means (see Fig. 8.4):

- The effective band gap of a quantum well is larger than E_{g} of the respective material.
- It can be tuned by the quantum well width L_z!

For optical inter band transitions between sub-band number n in the conduction band and sub-band number n' in the valence band, some selection rules must be obeyed:

- For infinite barrier heights, $\Delta n = n - n' = 0$.

- For finite barrier height (symmetric quantum wells), transitions are also allowed for $\Delta n = 0, 2, 4, \ldots$.
- The polarization of e-hh transitions is pure TE, while e-lh is mixed TM and TE.

Also possible: Inter-sub-band transitions in the same band (application: quantum cascade laser).

8.3 Density of States

Remember Section 1.10: The density of states (DOS) of bulk material is described by a square root law:

$$D(E) \sim \sqrt{E}$$

This was deduced by taking into account the three-dimensional properties of our crystal, i.e., by calculating the number of states within a sphere in the reciprocal space.

In a quantum well, the electrons can only move in the (x, y)-plane. Therefore, they are called **two-dimensional carriers** (for electrons: two-dimensional electron gas, sometimes abbreviated as TDEG).

When applying the same calculations for this two-dimensional case, the final result is

$$D(E) = \sum_{\mathrm{n}} D_{\mathrm{n}}(E) \quad \text{with} \quad D_{\mathrm{n}}(E) = \frac{g_s m^*}{2\pi\hbar^2}\Theta(E - \varepsilon_{\mathrm{n}}), \qquad (8.13)$$

where g_s = spin degeneracy factor (mostly = 2) and $\Theta(E)$ is the Heavyside step function:

$$\Theta(E) = \begin{cases} 0 & \text{for } E < 0 \\ 1 & \text{for } E > 0 \end{cases}$$

This means: In a single quantum well sub-band, the DOS is constant versus energy. In particular, it has this value already at the lowest energy of this sub-band, i.e., even for the lowest excitation energy of such bands, a finite (comparably large) number of states is available. When increasing the energy, we obtain a stepwise increasing DOS (Fig. 8.5) when higher sub-bands are taking part, whereas in the three-dimensional case, the DOS is zero at the lowest excitation

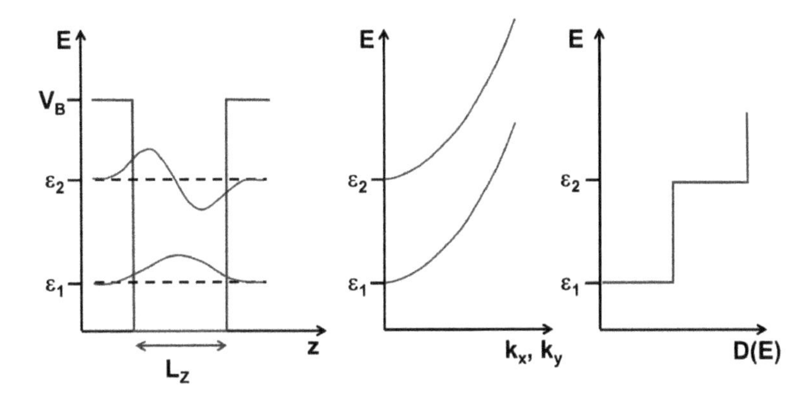

Figure 8.5 Wave function (left), dispersion relation in quantum well plane (middle) and density of states of the two first sub-bands of a quantum well (after [4]).

energy and increases only smoothly starting from $E = 0$. This quantum well behavior is excellent for energetically well defined optical transitions.

8.4 Behavior at $\vec{k} \neq 0$

When looking closer, then we realize that the parabolic dispersion relations (Eq. 8.4) are only a good approximation around $\vec{k} = 0$. Even here, the curvature (in the (x, y)-plane) changes, i.e., the effective masses change in a quantum well.

Moreover, an interaction of the different sub-bands leads to a more complex band structure (see Fig. 8.6).

To summarize up to here:

- Carriers are well confined in a quantum well due to the barriers. Remember: Radiative optical transitions increase with the carrier concentration n squared. Better carrier confinement means much more effective optical transitions (we can also say: much better overlap of electron and hole wave functions).
- The effective band gap depends on L_z, i.e., the emission wavelength of optoelectronic devices can be tuned just by the layer thickness without needing new materials!

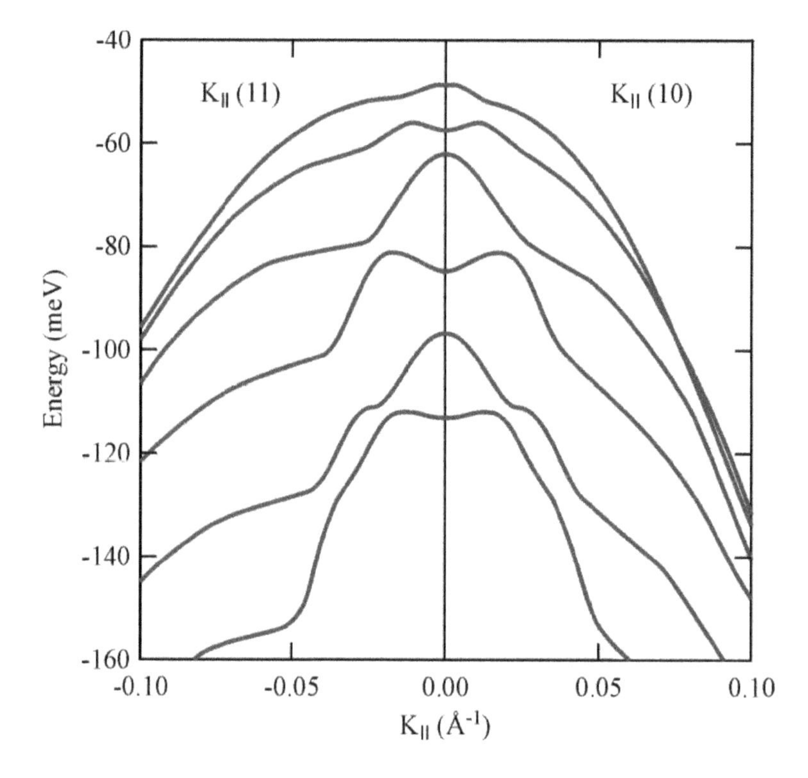

Figure 8.6 Valence band structure of a 12 nm GaAs quantum well (barrier height: 70 meV; after [4]), courtesy of A. Fily.

- The step-wise density of states further increases the number of carriers with same energy for recombination leading to a stronger and better defined light emission.
- The valence band degeneracy is lifted because of the different effective masses leading to different quantization.

8.5 High Carrier Mobility in a Quantum Well

Typically, carriers have higher mobilities in a quantum well which improves their properties for high frequency **electronic** devices like field-effect transistors. This is due to the following reasons:

- High carrier concentrations can be obtained by doping the barrier. The extra carriers will fall into the quantum well. Hence, impurities and carriers are locally separated. Respective scattering processes are strongly reduced.
- Other scattering fields may be screened by the comparably high carrier density in a quantum well.
- Do to the limitation on two dimensions, fewer final states for the scattering processes are available resulting in fewer scattering events!

These features are in particular used in **high electron mobility transistors** (HEMTs, see Chapter 13). They make use of the triangular quantum well which forms at the GaAs-AlGaAs interface (or at respective interfaces in other material combinations, see Fig. 2.4).

8.6 Coupling of Several Quantum Wells: Resonant Tunneling

As discussed in Section 8.2, the wave functions of the eigenstates of a quantum well with finite barriers penetrate into the barrier material. If the barriers between two or several quantum wells are thin enough, then the respective sub-bands of neighboring quantum wells couple (which may be regarded as coupled by the quantum-mechanical tunneling effect). This is completely analogous to the coupling of two LC resonance circuits or two atomic levels in a molecule, etc.: two sub-bands of the same energy form a coupled pair of sub-bands with some energy spacing depending on the coupling strength. In the case of many quantum wells, mini bands may form. This structure is often called a **superlattice**.

The sub-band position of a quantum well embedded by two thin barriers (Fig. 8.7) may be shifted in energy by the application of an electric field (external voltage). Then, the eigenstate may or may not be on the same energy level as the states in the adjacent material depending on the applied voltage. Carriers will only tunnel through these barriers if the respective sub-band lines up in energy. Hence, the electric transport through such a structure depends on

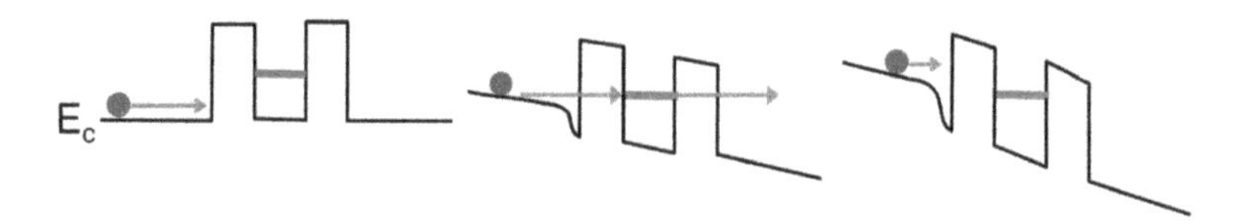

Figure 8.7 Conduction band structure of a resonant tunneling diode. Left: No applied voltage, no tunneling possible, because quantum well eigenstate is higher in energy than the conduction band edge outside. Center: With appropriate electric field, the eigenstate can be shifted to the same level as the conduction band edge on the left hand side. Now, electrons can tunnel into the quantum well and even further to the right hand side. Right: The eigenstate is too low to support tunneling, hence the current decreases again for further increasing applied voltage.

the applied voltage and may have negative differential resistance (resonant tunneling diode).

Problems

(1) Prove the formula about the quantum well energy states (Eqs. 8.6 and 8.7 of the script). Calculate the transition energies from the conduction band to the valence band for a GaAs-$Al_xGa_{1-x}As$ quantum well structure with $x = 0.25$, $L_z = 5$ nm. How many sub-bands exist in either band? Hint: Assume the same effective masses for GaAs and AlGaAs.

(2) In a quantum well, the degeneracy of the heavy and light hole bands at $k = 0$ is lifted. Explain. You may find two (fairly independent) arguments.

(3) (a) Explain: What makes the density of states (DOS) of a quantum well favorable as compared to the bulk DOS?

(b) Determine the number of states in the conduction band between the band edge E_C and E_C+40 meV. Do the same for a quantum well as given in the last problem from the energy of the lowest sub-band $E_{C,n1}$ to $E_{C,n1} + 40$ meV.

(4) Why is the carrier mobility in a quantum well typically larger than in bulk material? Consider room temperature and low temperature.

(5) (a) Explain the terminus "tunnel effect."

(b) Explain: What is a resonant tunneling diode? How can we understand its negative differential resistance region in the respective I–V curve?

(6) (a) What happens to the ground state of two quantum wells if they are separated only by a thin barrier? Find analogue problems in mechanics or electrical engineering.

(b) What happens if many quantum wells are arranged periodically as a superlattice with fairly thin barriers?

Chapter 9

Strain

Up to now, we focused on heterostructures where all single-layer materials have the same crystal structure and identical lattice constant. Then, perfect interfaces can be expected. However, when putting different materials epitaxially one on top of the other, this condition is not always perfectly fulfilled. In this chapter, we will discuss some consequences from lattice mismatch between layers in a heterostructure. In some cases, this leads to the development of **strain**, as discussed below. Originally, people tried to avoid such strain by controlling the compositions of the heterostructures accurately. Later, it turned out that strain may lead even to dramatic advantages in some applications.

Again, we only can discuss this topic briefly here. More details can be found in [49, 50].

Why is strain important?

- As mentioned above, it is a consequence of ("never perfect") hetero-epitaxy. Therefore, it is present in larger or smaller amount in nearly every epitaxial structure.
- It has a strong influence on the band structure of the materials (via the displacement of the atoms). Hence, it is a tool to intentionally change the band structure.

Compound Semiconductors: Physics, Technology, and Device Concepts
Ferdinand Scholz

ISBN 978-981-4774-07-9 (Hardcover), 978-1-315-22931-7 (eBook)
www.panstanford.com

What causes the development of strain? If a layer is grown epitaxially on another (single crystal) material, then the atoms of the new layer arrange themselves with respect to the atoms of the "substrate." This even happens for a material with a lattice constant slightly different than the substrate: The atoms are forced to keep the in-plane lattice structure, i.e., the in-plane lattice constant:

$$a_{||}(\text{layer}) = a_{||}(\text{substrate})$$

even if the equilibrium (or "relaxed") lattice constants a_r of the two materials do not match.[a] This means that the in-plane lattice constant of the layer is changed by the invoked strain:

$$a_{||}(\text{layer}) \neq a_{\mathrm{r}}(\text{layer}),$$

which can be described with the respective displacement or **strain**

$$\varepsilon_{||} = \left.\frac{\Delta a}{a}\right|_{||} = \frac{a_{||} - a_{\mathrm{r}}}{a_{\mathrm{r}}}.$$

This situation is called **pseudomorphic strain**, because there is a perfect match of all crystalline bonds at the interface.

Another reason for such "in-plane strain" or **biaxial strain** may be the different thermal expansion (see Section 1.8) of two adjacent layers in a heterostructure, as is well known from bimetal temperature sensors. Such thermally induced strain typically develops when a heterostructure cools down after the epitaxial process.

Typical values for such strain are about 1% for lattice mismatch induced strain (see Section 9.3) and 0.1% for thermally induced strain.

In both cases, the biaxial strain $\varepsilon_{||}$ also leads to a displacement $\varepsilon_{\perp}$ in the vertical direction (Fig. 9.1). The determination of $\varepsilon_{\perp}$ requires a short excursion to continuum mechanics.

9.1 Elastic Deformation of a Solid

In a solid, forces (stress $\underline{\sigma}$) and displacements (strain $\underline{\varepsilon}$) are related by **Hooke's law**, as long as stress and strain are not too large. This means: Stress and strain are linearly related:

$$\sigma_{ij} = C_{ijkl}\varepsilon_{kl} \tag{9.1}$$

---

[a] In the following, we assume for simplicity that the substrate's lattice constants do not change. This is allowed because of the large thickness of the substrate.

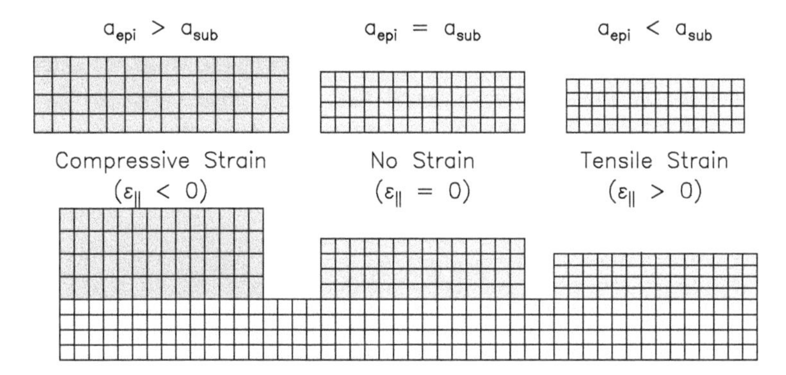

Figure 9.1 Pseudomorphic strain of a lattice mismatched layer (schematically).

According to the generally anisotropic nature of solids, $\underline{\sigma}$ and $\underline{\varepsilon}$ are 3×3 tensors of second order, connected by $\underline{\underline{C}}$, a tensor of fourth order, which contains the material specific elastic constants.

On a first glance, this tensor has $3^4 = 81$ components. However, due to symmetry constraints and as we do not use this law for describing, e.g., a rotation of the solid, just 36 remain, as $\underline{\sigma}$ and $\underline{\varepsilon}$ have only 6 remaining components each.

From thermodynamics, we know that the strain energy density e reads

$$e = \frac{1}{2}\underline{\sigma} \cdot \underline{\varepsilon} = \frac{1}{2}\underline{\varepsilon} \cdot \underline{\underline{C}} \cdot \underline{\varepsilon}$$

which means (as a consequence of the symmetry of the latter part of this equation) that only 21 components remain.

In the general case of a perfectly anisotropic solid (21 remaining elastic constants), it is possible to simplify the writing of these calculations by using **Voigt's notation**: The tensors $\underline{\sigma}$ and $\underline{\varepsilon}$ are now written as 6-component vectors, and the second order tensor $\underline{\underline{C}}$ (3^4 components) is taken as a first order tensor with only $6 \cdot 6$ components (and $C_{ij} = C_{ji}$). Then Hooke's law reads

$$\begin{pmatrix} \sigma_{xx} \\ \sigma_{yy} \\ \sigma_{zz} \\ \sigma_{xy} \\ \sigma_{yz} \\ \sigma_{zx} \end{pmatrix} = \begin{pmatrix} C_{11} & C_{12} & C_{13} & C_{14} & C_{15} & C_{16} \\ C_{12} & C_{22} & C_{23} & C_{24} & C_{25} & C_{26} \\ C_{13} & C_{23} & C_{33} & C_{34} & C_{35} & C_{36} \\ C_{14} & C_{24} & C_{34} & C_{44} & C_{45} & C_{46} \\ C_{15} & C_{25} & C_{35} & C_{45} & C_{55} & C_{56} \\ C_{16} & C_{26} & C_{36} & C_{46} & C_{56} & C_{66} \end{pmatrix} \cdot \left(\varepsilon_{xx}, \varepsilon_{yy}, \varepsilon_{zz}, \varepsilon_{xy}, \varepsilon_{yz}, \varepsilon_{zx}\right).$$

This number of 21 components may be further reduced by the symmetry of the crystal. A perfectly isotropic solid material, for example amorphous material like glass, possesses only two components, the bulk modulus K and the shear modulus G (or the two Lamé constants λ and μ). Then Hooke's law reads

$$\begin{aligned}\underline{\sigma} &= K \cdot \varepsilon_I \cdot \underline{I} + 2G \cdot \underline{\varepsilon}' \\ &= \left(K - \frac{2}{3}G\right)\varepsilon_I \cdot \underline{I} + 2G \cdot \underline{\varepsilon} \\ &= \lambda \cdot \varepsilon_I \cdot \underline{I} + 2\mu \cdot \underline{\varepsilon} \\ \text{with} \quad \varepsilon_I &= \varepsilon_{xx} + \varepsilon_{yy} + \varepsilon_{zz} \\ \text{and} \quad \underline{\varepsilon}' &= \underline{\varepsilon} - \frac{1}{3}\varepsilon_I \cdot \underline{I}.\end{aligned}$$

Another constant often used in such calculations is Poisson's number:

$$\nu = -\frac{\varepsilon_{||}}{\varepsilon_{\perp}} = \frac{\lambda}{2(\lambda + \mu)}$$

In a cubic crystal with some anisotropy as compared to amorphous materials, only 3 components remain, and Hooke's law then reads

$$\begin{pmatrix}\sigma_{xx}\\ \sigma_{yy}\\ \sigma_{zz}\\ \sigma_{xy}\\ \sigma_{yz}\\ \sigma_{zx}\end{pmatrix} = \begin{pmatrix}C_{11} & C_{12} & C_{12} & 0 & 0 & 0\\ C_{12} & C_{11} & C_{12} & 0 & 0 & 0\\ C_{12} & C_{12} & C_{11} & 0 & 0 & 0\\ 0 & 0 & 0 & C_{44} & 0 & 0\\ 0 & 0 & 0 & 0 & C_{44} & 0\\ 0 & 0 & 0 & 0 & 0 & C_{44}\end{pmatrix} \cdot \left(\varepsilon_{xx}, \varepsilon_{yy}, \varepsilon_{zz}, \varepsilon_{xy}, \varepsilon_{yz}, \varepsilon_{zx}\right).$$

Now let us consider biaxial strain in a cubic crystal (like GaAs) in the $\{001\}$ plane: Here, the atoms are subjected to some in-plane force (stress) $\sigma_{xx} = \sigma_{yy} \neq 0$, whereas there is no force perpendicular to the surface: $\sigma_{zz} = 0$ ("free-floating" surface).

There are no shear forces; therefore all components mixed in x, y, and z are zero.

Hence, for biaxial strain in a $\{001\}$ plane, the following equations are valid:

$$\begin{aligned}\sigma_{xx} &= C_{11} \cdot \varepsilon_{xx} + C_{12} \cdot \varepsilon_{yy} + C_{12} \cdot \varepsilon_{zz} \\ \sigma_{yy} &= C_{12} \cdot \varepsilon_{xx} + C_{11} \cdot \varepsilon_{yy} + C_{12} \cdot \varepsilon_{zz}\end{aligned}$$

$$\sigma_{zz} = C_{12} \cdot \varepsilon_{xx} + C_{12} \cdot \varepsilon_{yy} + C_{11} \cdot \varepsilon_{zz} = 0$$

the latter because there is no force in z-direction.

From the last equation, it follows that

$$C_{12} \cdot \varepsilon_{xx} + C_{12} \cdot \varepsilon_{yy} = -C_{11} \cdot \varepsilon_{zz}.$$

Hence, we obtain a simple relation between the in-plane strain $\varepsilon_{xx} = \varepsilon_{yy} = \varepsilon_{||}$ (they are equal because of the symmetry of a cubic crystal) and the out-of-plane strain $\varepsilon_{zz} = \varepsilon_{\perp}$:

$$\varepsilon_{\perp} = -\frac{2C_{12}}{C_{11}}\varepsilon_{||} \tag{9.2}$$

This means: By knowing Hooke's constants C_{ij} of the materials, we can predict how the out-of-plane lattice constant $a_{\perp}$ changes for a given pseudomorphic strain.

If the biaxial strain is in other crystal planes, then other relations are valid.

9.2 Influence on Electronic Band Structure

As qualitatively discussed in Chapter 1, the details of the band structure depend on the crystalline structure, i.e., on the position of the atoms. Hence, strain does influence the band structure.

Let us first look to the fundamental properties, i.e., the band edge positions at $\vec{k} = 0$, which determine the band gap in direct semiconductors. The following description goes along publications from Krijn [51] and van de Walle [52] for zinc blende crystals, biaxially strained in the $\{001\}$ plane (the most common epitaxial direction for III-V compounds besides GaN).

The biaxial strain can be separated into a hydrostatic part (isotropic change of volume) and in a part containing only the shear strain.

The hydrostatic part only acts on the energy of the conduction band edge E_{c}

$$\Delta E_{\mathrm{c}} = a_{\mathrm{c}}(2\epsilon_{||} + \epsilon_{\perp}), \tag{9.3}$$

and in the same way on the mean valence band edge $E_{\mathrm{v,av}}$, i.e., the average of heavy- and light-hole and spin-orbit-split-off bands:

$$\Delta E_{\mathrm{v,av}} = a_{\mathrm{v}}(2\epsilon_{||} + \epsilon_{\perp}), \tag{9.4}$$

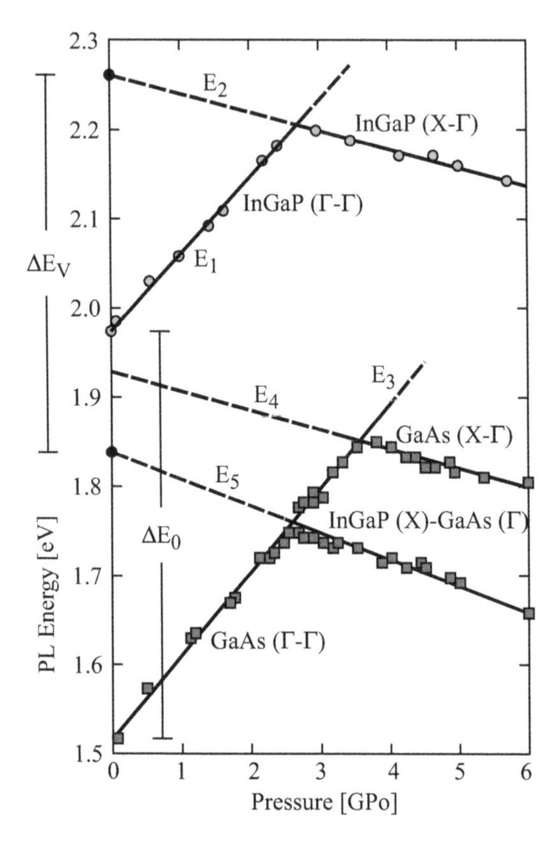

Figure 9.2 Photoluminescence transition energies as a function of hydrostatic pressure for GaInP (circles) and GaInP-GaAs multi-quantum well structure (squares). For higher pressures, conduction band minima at $k \neq 0$ become the absolute minima making the band structure indirect. Reprinted from [53], with the permission of AIP Publishing.

where a_c and a_v are the hydrostatic deformation potentials of conduction and valence band, respectively.

Notice: Both band edges typically move in opposite direction leading to a larger band gap for compressive and a smaller one for tensile strain.[b] The hydrostatic deformation potentials of the various bands or band extrema may be very different and even have opposite

[b]Notice the same trend when looking to the band gap of various materials where those with larger lattice constant have a smaller band gap (see Fig. 2.3).

sign (Fig. 9.2). In the latter case, this may give rise to a transition from a direct to an indirect band structure with increasing strain.

The shear component causes a symmetry reduction of the lattice. This influences just the valence band. As discussed for quantum wells, this effect lifts the valence band degeneracy of heavy and light holes at $\vec{k} = 0$, as this kind of anisotropic strain breaks the symmetry. Now, the valence band edges are shifted with respect to the original position of the heavy hole band edge $E_{\text{hh (without shear strain)}} = E_{\text{v,av}} + \Delta_0/3$ by

$$\Delta E_{\text{hh}} = -\frac{1}{2}\delta E \tag{9.5}$$

for the heavy holes and by

$$\Delta E_{\text{lh}} = -\frac{1}{2}\Delta_0 + \frac{1}{4}\delta E + \frac{1}{2}\left[(\Delta_0)^2 + \Delta_0\delta E + \frac{9}{4}(\delta E)^2\right]^{\frac{1}{2}} \tag{9.6}$$

for the light holes. In this formula,

$$\delta E = 2b(\epsilon_{\perp} - \epsilon_{\parallel})$$

is the strain-dependent shift with the shear deformation potential b and Δ_0 is the split-off energy without strain.

A third somewhat more complex formula is valid for the shift of the split-off valence band.

The conduction bands are not influenced by the shear strain due to their s-like character (spherical symmetry).

Finally, the transition energies of electrons from the conduction band edge to the valence band edges can be deduced:

$$\Delta E_{e,ih} = E_g + \Delta E_c - \Delta E_{v,av} - \Delta E_{ih}, \tag{9.7}$$

where the index i stands for the different valence bands.

Figure 9.3 shows these energies versus different hydrostatic and biaxial strain for GaAs as an example. Notice: In compressively strained material ($\varepsilon_{||} < 0$), the heavy-hole valence band is the top-most band, whereas it is the light-hole band for tensile strain. In both cases, the band edge E_g is significantly changed.

In practice, the biaxial strain is often varied by changing the composition and hence the lattice constant a_{layer} of a ternary layer grown on top of a material with given lattice constant $a_{\text{substrate}}$. Thus,

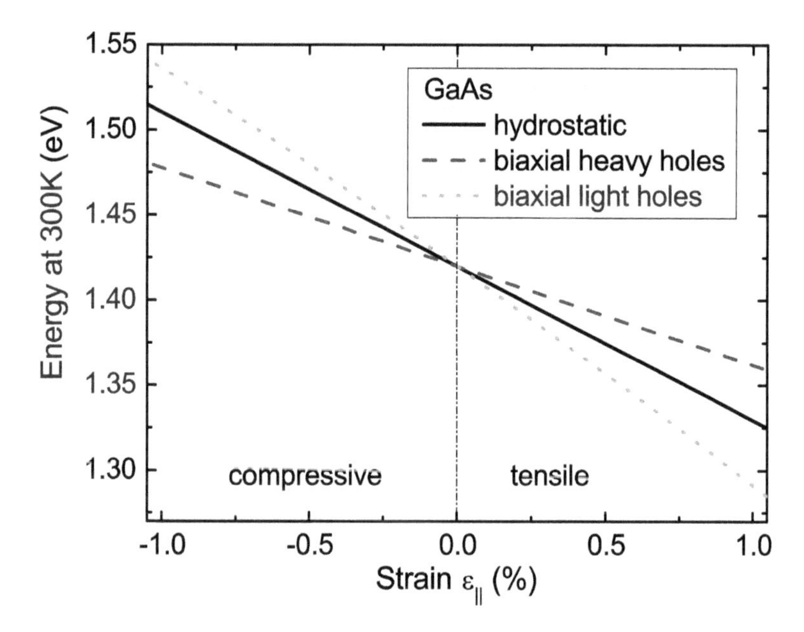

Figure 9.3 Band edge transition energies of biaxially strained GaAs at room temperature. Full line: Only hydrostatic pressure. Broken lines: Biaxial strain, heavy and light holes.

the finally visible band gap depends additionally on the fundamental band gap of this layer according to Eq. 2.2 (see Fig. 9.4).

Besides the energies at $\vec{k} = 0$, also the dispersion relations, hence the curvatures and consequently the effective masses are influenced by strain. There is only little influence on the curvature of the conduction band. However, the valence bands see a drastic change: They get a completely anisotropic character. If we neglect the spin-orbit coupling, then also the masses vertical to the biaxial plane remain unchanged [54, 55]:

$$m_{hh,\perp} = \frac{1}{\gamma_1 - 2\gamma_2} = m_{hh} \tag{9.8}$$

$$m_{lh,\perp} = \frac{1}{\gamma_1 + 2\gamma_2} = m_{lh}, \tag{9.9}$$

where γ_i are the Luttinger valence band parameters [56, 57].

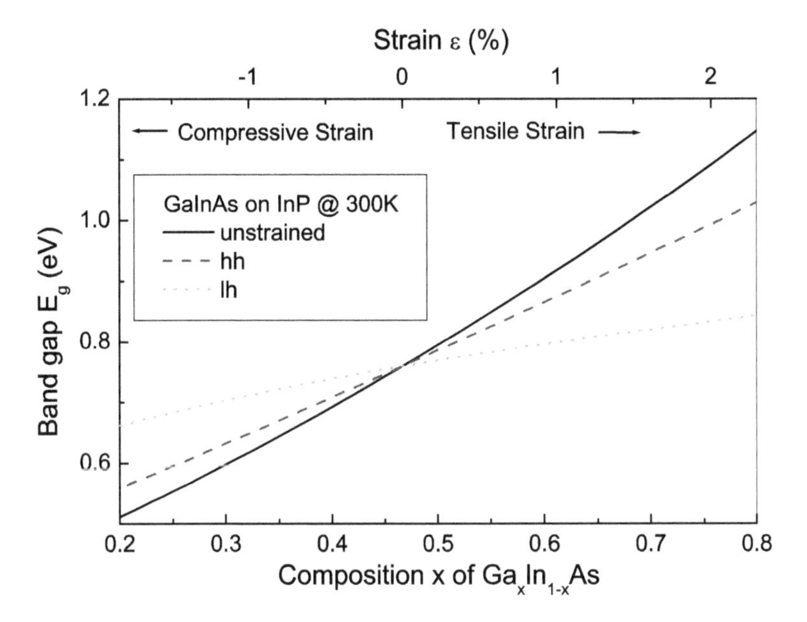

Figure 9.4 Transition energies from the conduction band edge to the valence band edges of the heavy and light hole bands in $Ga_xIn_{1-x}As$ on InP. The top axis shows the respective strain for full pseudomorphic growth is which assumed here.

In the plane, for small values of $\vec{k}$, the masses change according to [54]:

$$m_{hh,\parallel} = \frac{1}{\gamma_1 + \gamma_2} \tag{9.10}$$

$$m_{lh,\parallel} = \frac{1}{\gamma_1 - \gamma_2}. \tag{9.11}$$

Please notice: In the plane,

- the originally heavy holes get lighter and
- the originally light holes get heavier

as illustrated in Fig. 9.5. In either case, the top-most holes which define the band gap get lighter than in the unstrained case. This has big consequences for devices where the carrier mobility (high frequency electronics) or the density of states (laser diodes) is important.

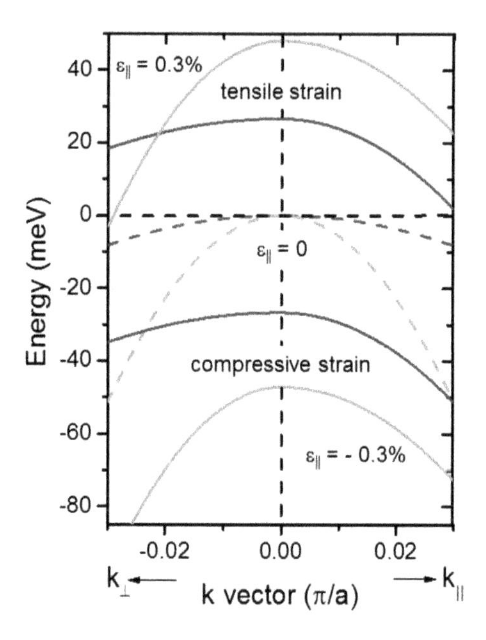

Figure 9.5 Dispersion relations (schematically) of the valence bands in case of biaxial tensile (upper curves) and compressive (lower curves) strain for GaAs (full lines); for comparison, the dispersion relations of unstrained material are indicated as broken lines (middle curves). The energy values (left scale) are based on the data from [51], taking the maxima of the unstrained valence bands as reference. For simplicity, parabolic curvatures are assumed for these bands.

9.3 Critical Thickness

When a lattice-mismatched layer grows epitaxially and pseudomorphically strained on a substrate, then it is clear that the strain energy in the layer increases with thickness. Finally, this strain energy may surpass some critical value, and the strain can no longer be maintained. Instead, crystalline defects (dislocations) are generated which accommodate the lattice mismatch.

This can be calculated by evaluating the equilibrium of energy or of forces

- which result from the strain;
- which are necessary to create a defect.

An often cited formula for the critical thickness h_c, where this relaxation starts, was determined by Matthews and Blakeslee [58]:

$$h_c = \frac{b}{2\pi\varepsilon}\frac{1-\nu\cos^2\alpha}{(1+\nu)\cos\lambda}\left(\ln\frac{h_c}{b}+1\right) \tag{9.12}$$

where ν = Poisson's ratio, b = the relevant Burger's vector of the involved dislocations, α = the angle between $\vec{b}$ and the dislocation, λ = angle between slip direction and direction in film plane perpendicular to the line of intersection of the slip plane and the interface, and ε = the mismatch strain.

The result is plotted in Fig. 9.6 for a typical compound semiconductor.

We notice that larger strain (in the range of 1%) can only form in thin layers, i.e., quantum wells.

This means: Typically we have to consider both the influence of **strain** and **quantization** simultaneously. This further complicates our band structure calculations.

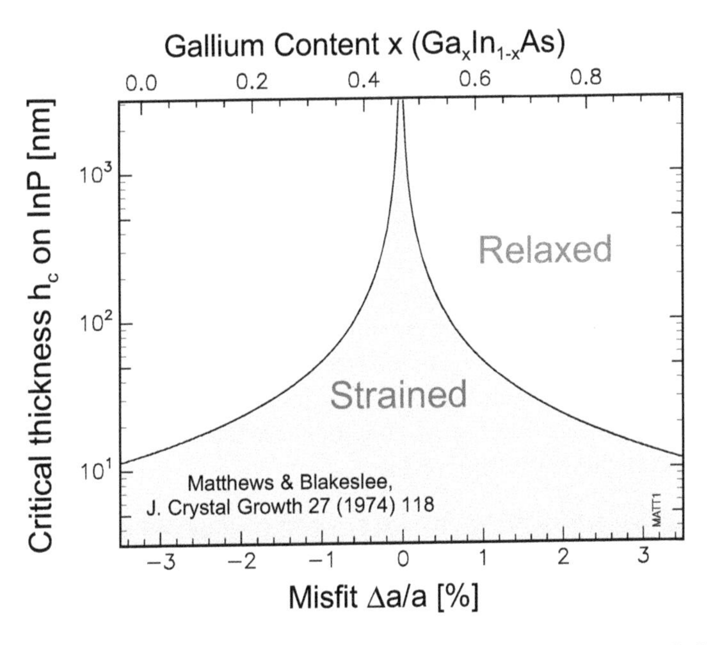

Figure 9.6 Critical thickness for a $Ga_xIn_{1-x}As$ layer grown on InP (as calculated by Eq. 9.12).

In some cases, they even counteract each other. Remember: For tensile strain, the heavy-hole band is shifted downward stronger than the light-hole band. Quantization always acts stronger on the light-hole band. The resulting crossing of the bands leads to a fairly complex band structure where even a maximum at $k \neq 0$ may form in the valence band transforming the respective quantum well into an **indirect** semiconductor!

9.4 Usefulness of Strain

Why is it useful anyway? In a simple picture, strain reduces the effective hole mass of the top-most valence band. This is quite advantageous for, e.g., a laser diode: The maximum gain γ_{max} of a quantum well is given by [4]

$$\gamma_{\mathrm{max}} = \alpha_{2D}(1 - e^{-\frac{n_s}{n_C}} - e^{-\frac{n_s}{n_C}\frac{m_e}{m_h}}).$$

Here, α_{2D} is the absorption coefficient of the quantum well, $n_s = p_s = \frac{j\tau_{\mathrm{tot}}}{q}$ is the carrier density, and n_C is the critical two-dimensional density ($n_C = D(E) \cdot kT$).

Transparency is reached when γ_{max} gets positive, this happens best for $m_e/m_h \simeq 1$. In a strained quantum well, the hole mass m_h is reduced, hence closer to m_e.

Indeed, the implementation of strain into semiconductor laser structures was another big step towards lower threshold currents.

In order to get larger strain or more quantum wells with a given strain, the concept of **strain compensation** is applied: The barriers are strained just in the opposite direction as the quantum wells.

Depending on the sign of the strain (compressive or tensile), the (formerly) heavy-hole band or light-hole band builds the valence band edge, respectively. In consequence, the polarization of the emitted light may be different:

- TE polarization[c] (as in most cases) for compressive strain and
- TM polarization for tensile strain.

[c]TE = transverse electric polarization: The electric field of the light traveling along the epitaxial structure (in a waveguide built by some particular epitaxial layers) oscillates parallel to the layer planes. In TM = transverse magnetic polarization, the electric field oscillates perpendicular to the layer planes.

Waveguides formed by the semiconductor heterostructure (see later) favor TE polarization. This makes tensile strained quantum wells somewhat less favorable for laser applications (besides the problem of the band crossing at slight tensile strain).

Problems

(1) Find the thermal expansion coefficients of GaAs and Si. Then determine the strain of a thin GaAs layer at room temperature on a thick Si substrate. Assume that GaAs is epitaxially grown on the Si substrate at a temperature of 700°C as unstrained layer and gets biaxially strained during cool-down. Calculate the band gap of the strained GaAs.
Use the following GaAs material constants:

Poisson ratio	$\frac{2C_{12}}{C_{11}}$	0.92	
Split-off energy	Δ_0	0.34	eV
Hydrostatic deformation potential of valence band	a_V	1.16	eV
Hydrostatic deformation potential of conduction band	a_C	−7.17	eV
Shear deformation potential	b	−1.7	eV

(2) (a) Explain: How does a layer get biaxially strained when grown on a substrate?
(b) Explain the termini and indicate their units:

- strain
- stress
- lattice mismatch

(3) In basic physics lectures, we learn that the deformation of solid material can be described with 2 elastic coefficients. Under which conditions is this statement correct? When do we need 3, 5, or 21 elastic coefficients?

(4) (a) Deduce the ratio of $\varepsilon_{\perp}$ and $\varepsilon_{||}$ for a cubic crystal for biaxial strain in the (001) plane.

(b) What is "uni-axial strain"?
(c) Consider uni-axial strain applied along the <001> direction of GaAs. How is now the ratio of $\varepsilon_{\perp}$ and $\varepsilon_{||}$?

(5) How does the band gap change with

- hydrostatic pressure;
- biaxial strain?

Find some simple argument by looking at Fig. 2.3.

(6) (a) Explain the term "critical thickness" with relation to strained epitaxial layers.
(b) How could you measure whether the thickness of a layer is below or above the critical thickness?

Chapter 10

Low-Dimensional Systems

From Chapter 8, we got the main message: Quantum wells help to substantially improve many optoelectronic and electronic devices. A major factor is their step-like density of states.

Hence the question is obvious: How about further reduction of dimensionality, i.e., limitation of the size of our object to values in the range of the de-Broglie wavelength also for the other space coordinates? If carriers are confined to just a line, we call the respective structure a "quantum wire." If it is confined in all three space dimensions, i.e., to a local point, then it is a "quantum box" or a "quantum dot." We will discuss later in this chapter how to realize such structures. Starting here with the theoretical expectations, we can easily see: Such reduced dimensionality is indeed very promising, as discussed below, and hence has been and still is a major focus of semiconductor research. There are many nice reviews available, including references [59–61].

10.1 Band Structure, etc.

In quantum wells, the carriers are confined in a plane. This situation could be calculated by solving the one-dimensional

Compound Semiconductors: Physics, Technology, and Device Concepts
Ferdinand Scholz

ISBN 978-981-4774-07-9 (Hardcover), 978-1-315-22931-7 (eBook)
www.panstanford.com

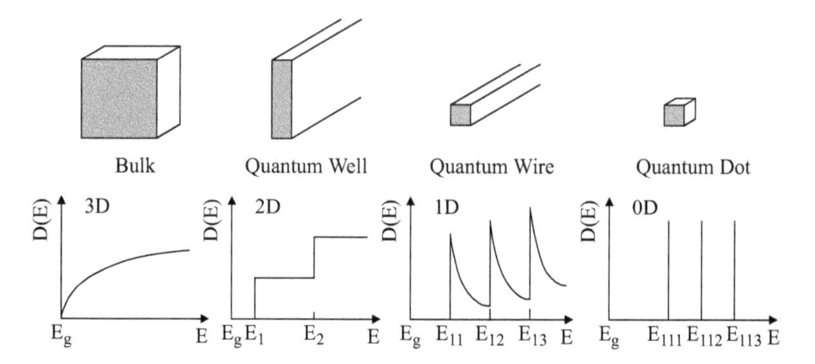

Figure 10.1 Density of states $D(E)$ (schematically) for decreasing dimensionality.

quantum mechanical problem of a "particle in a box." Similarly, lower-dimensional systems can be treated by solving this problem now for two-dimensional (quantum wires) or three-dimensional confinement (quantum dots). Here, we do not go into the details of the Schrödinger equation but only briefly explain the most important results.

As expected, one of the most significant changes while reducing the dimension of the possible carrier movement concerns the density of states (DOS), as indicated here (see Fig. 10.1):

For three dimensions (bulk), we found:

$$D_{3D}(E) \sim \sqrt{(E - E_g)} \tag{10.1}$$

whereas for a quantum well, we obtained:

$$D_{2D}(E) \sim \sum_m \Theta(E - E_m) \tag{10.2}$$

Similarly, for a quantum wire we get:

$$D_{1D}(E) \sim \sum_{l,m} \frac{1}{\sqrt{E - E_{l,m}}} \tag{10.3}$$

and eventually for a quantum dot:

$$D_{0D}(E) \sim \sum_{l,m,n} \delta(E - E_{l,m,n}) \tag{10.4}$$

with $\Theta(E)$ being the Heavyside step function and $\delta(E)$ representing Dirac's pulse function.

Hence, the DOS gets more and more favorable for, e.g., laser applications, as the optical gain is given by (c.f. Eq. 6.2)

$$g_{(i)}(E) = \frac{\pi^2 c^2 \hbar^3}{\bar{n}^2 E^2} B_{(i)} \int_0^{E-E_g} D_{c(i)}(\varepsilon) D_{v(i)}(\varepsilon - E)[f_c(\varepsilon) f_v(\varepsilon - E)] d\varepsilon, \tag{10.5}$$

where $\bar{n}$ is the refractive index.

We can see that due to the higher DOS at the lowest energy level available in such systems, the optical gain at this specific energy is increasing in lower dimensions (Fig. 10.2 left).

For transparency (i.e., when the optical gain as a consequence of stimulated emission compensates all optical losses, see Section 12.2), the necessary carrier density is about the same for all dimensions. Hence the absolute number of required carriers and thus the current density in a laser diode decreases.

Moreover, the differential gain $A(n) = \frac{\partial g}{\partial n}$ increases, which also goes into the modulation properties of a laser diode. Its maximum modulation frequency is directly related to the differential gain by

$$f_r \sim \sqrt{A(n)} = \sqrt{\frac{\partial g}{\partial n}} \tag{10.6}$$

Finally, the temperature dependence for the optical gain decreases with decreasing dimensionality (Fig. 10.2 right). The effective density of states N depends on temperature as $T^{d/2}$ where d is the dimensionality, hence

$$\begin{aligned} N(3\mathrm{D}) &\sim T^{3/2} \\ N(2\mathrm{D}) &\sim T \\ N(1\mathrm{D}) &\sim T^{1/2} \\ N(0\mathrm{D}) &\sim T^0 = \text{const.} \end{aligned}$$

Thus, no temperature dependence is expected for a quantum dot (note the δ-like density of states). Indeed, a quantum dot is very similar to a single atom where the electrons are confined to the atomic nucleus. Now, it is no surprise that atoms have very sharp spectral lines, as the density of states of these electrons is also δ-like.

For such low-dimensional structures, quantization occurs similarly as discussed for a quantum well. If we assume, as for a quantum well, rectangular boundaries (i.e., a quantum dot is taken as a cuboid with dimensions L_x, L_y, L_z), then we can evaluate eigenvalues $E_{\mathrm{l,m,n}}$

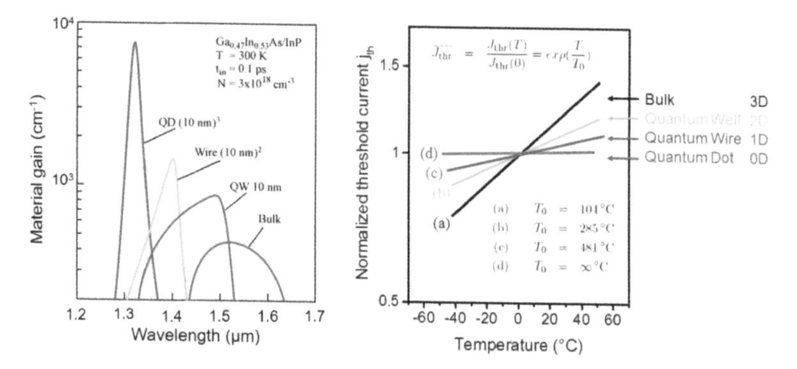

Figure 10.2 Expected gain (left) and temperature dependence of the laser threshold (right) for reduced dimensionality. Left diagram: © 1986 IEEE. Reprinted, with permission, from [62]. Right diagram: Reprinted from [63], with the permission of AIP Publishing.

for each direction similar as we did for the rectangular potential well in Section 8.2. In this special case, the Schrödinger equation can be separated by a product ansatz (c.f. Eq. 8.3):

$$H\Psi = E\Psi \quad \text{with} \quad \Psi = \Psi_x \cdot \Psi_y \cdot \Psi_z \tag{10.7}$$

Hence, the quantization can be calculated for each direction independently.

Another simple geometry for a quantum dot is just a sphere, which also can be treated theoretically quite easily.

Interestingly, a quantum dot has a bound state only for some minimum size and/or some minimum potential. For a spherical dot with radius R_0, this minimum potential is

$$V_{0,min} \geq \frac{\pi^2 \hbar^2}{8m^* R_0^2}.$$

For a smaller radius R, no bound state exists in this quantum dot. Remember: A (symmetric) quantum well has **always** at least one bound state, no matter how small its width L_z is.

A problem, which was heavily discussed for low-dimensional structures, in particular quantum dots (or quantum boxes), is the so-called **phonon bottleneck**:

After excitation, the carriers are typically deep in the bands, i.e., far above the band edge, and should relax fast into the respective ground levels of the conduction and valence bands. This process is enabled by the interaction with phonons which help to conserve both, energy and momentum. However, in a zero-dimensional quantum dot, the electrons cannot change their momentum (as they cannot move in either direction). Hence, it is unclear whether they find appropriate phonons for the relaxation process.

However, this bottleneck could not be observed experimentally. Possible reasons are

- multi-(acoustic) phonon relaxation;
- Auger-like processes;
- defect-related relaxation.

10.2 Fabrication of Low-Dimensional Structures

(1) Top-down approach: Wires or dots defined by lithography

This kind of approach describes methods where an artificial structurization including some lithographic steps is performed. This process typically starts with a quantum well which can be easily grown by epitaxy (c.f. Chapters 4 and 8). The following steps lead to a lateral structurization of the quantum well resulting in quantum wires and quantum boxes.

- *Lithography:* For the fabrication of low-dimensional structures, optical lithography is not well suited, because its resolution is limited by the wavelength of light (some hundreds of nanometers), whereas for electronic low-dimensional structures, dimensions of the order of the de-Broglie wavelength

$$\lambda = \frac{h}{p} = \frac{h}{\sqrt{2mE}}$$

must be realized (a few to some tens of nanometers). This can be obtained with electron beam lithography: Similar as in a scanning electron microscope, an electron beam is focused on the sample (covered with some photo resist). The required pattern is written by the electron beam which exposes the photo resist locally. The resolution can be driven below 10 nm.

- After developing the photo resist, the pattern may be transferred into the semiconductor by *dry etching*. This requires an anisotropic etching process which etches in vertical direction but with no or little lateral etching. This is typically a kind of ion bombardment of the surface.
  Fundamental problem: The ion bombardment produces crystalline defects at the side walls of the etch crater giving rise to a "dead zone" of amorphous material and structural defects. The width of this zone is in the range of some 10 nm. Hence small quantum structures only consist of such defective material. This leads to a drastic decrease of optical properties of such small structures.
  Some improvement may be obtained by
  - regrowth over the etched structure thus burying the quantum structures into material of higher band gap, and
  - wet etch dip to remove the dead zone (requiring excellent control of all steps to finally achieve the correct size of the structures).
- Other method: *Intermixing* of quantum well by defect-induced interdiffusion: Using the patterned photoresist as a mask, local defects can be created by ion implantation. In a subsequent annealing step, interdiffusion of, e.g., Al and Ga in a GaAs-AlGaAs quantum well structure leads to a local band gap increase and thus to the definition of quantum structures.
  However, here, too, a lot of defects are created even in the masked areas by scattering effects of the implanted ions.
  Although a lot of research went into such methods, there has not been really a breakthrough until today.
- Obviously, the structurization of a quantum well by etching or else (which is a kind of local destruction of the quantum well) creates a lot of defects. How about fabrication of low-dimensional structures directly by epitaxial growth?
  One possible approach:
  - Etch structures into some semiconductor material. By wet etching, fairly large structures (some micrometers

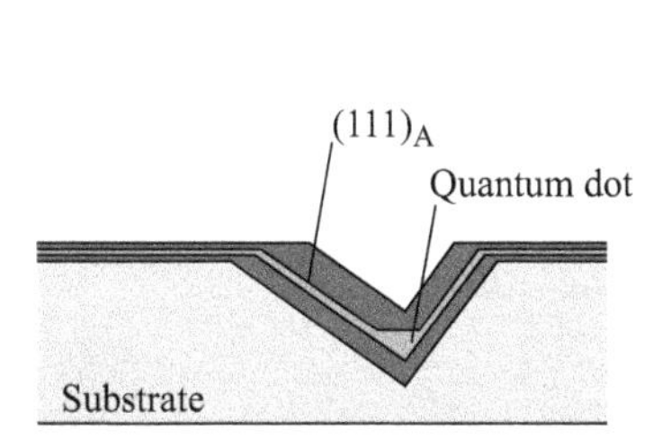

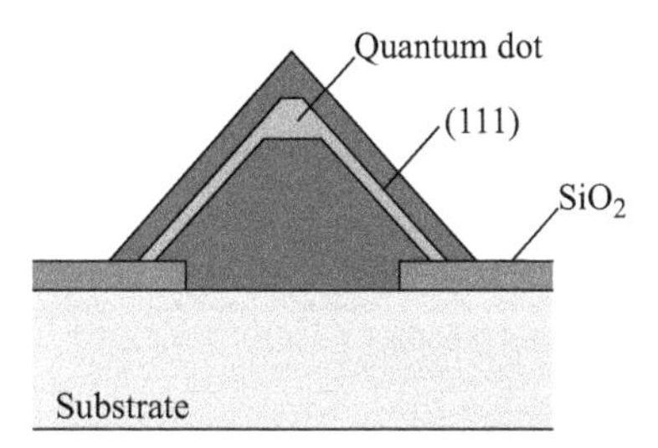

Figure 10.3 Growth of a quantum dot in V-shaped groove (inverted pyramid, left) or on top of a selectively grown pyramid (right). Quantum wires can be grown in the same way into stripe-like grooves or on top of stripes.

$\rightarrow$ optical lithography is o.k.) are etched into, e.g., GaAs. Due to the different chemical stability of different crystal facets, triangular stripe trenches with a very sharp tip can be easily etched (sidewalls formed by $\{111\}_A$ planes).

- Grow epitaxially a quantum well structure into such trenches. Due to the different surface migration properties of the different atoms (Al and Ga), overgrowth with AlGaAs leads to a further shapening of the triangular cross section, whereas GaAs will migrate down to the tip of the trench hence forming a quantum wire there (see Fig. 10.3, left).
- Due to the reshapening effect of AlGaAs, several quantum wires can be stacked (Fig. 10.4).

This kind of process has also been applied to the formation of quantum dots by first etching inverse pyramids.

The resulting quantum wires and dots have very high crystalline quality and could be successfully incorporated into the active zone of laser diodes.

Another similar approach starts with masking a semiconductor surface with a dielectric mask (e.g., SiO_2) containing some small openings (again produced by conventional optical lithography). In a subsequent epitaxial process, the growth only starts in these openings: Stripes or pyramids are formed by selective area epitaxy. Their shape depends

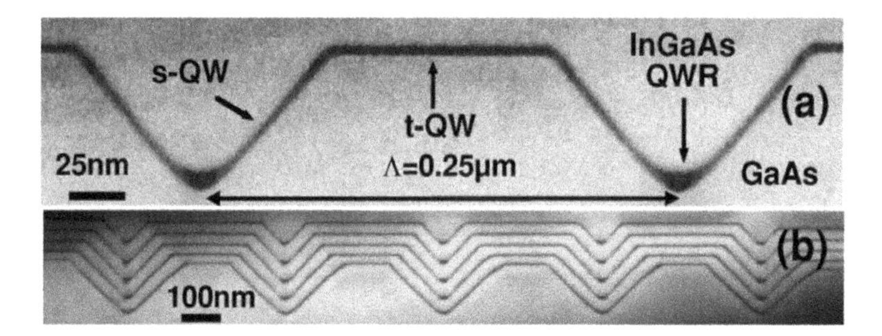

Figure 10.4 Transmission electron microscopy image of GaInAs quantum wires grown into V-grooves of GaAs by MOVPE (reprinted from [64], with the permission of AIP Publishing).

on the details of the growth conditions. Again very sharp tips can be realized. When depositing eventually another material, surface migration may lead to a quantum wire or quantum dot like structure at the tip of the stripe or pyramid (see Fig. 10.3, right). However, this process was not yet so successful, as the control of the quantum wire or quantum dot shape depends critically on the timing of the growth process.

(2) **Bottom-up approach: Self-assembled quantum dots**
In the last 20–25 years, quite a different approach has been developed for the realization of high quality quantum dots.
The basic idea is a kind of island formation during epitaxial growth. This growth mode is quite opposite to the two-dimensional growth which leads to good quantum wells and abrupt interfaces (c.f. Figs. 4.19 and 4.20).
In general, the specific growth behavior is governed by the energy balance at the surface during epitaxial growth:

$$\Delta\sigma = \sigma_{\mathrm{epi}} - \sigma_{\mathrm{substrate}} + \sigma_{\mathrm{interface}}, \tag{10.8}$$

where σ_{epi} and $\sigma_{\mathrm{substrate}}$ refer to the specific free surface enthalpies of the epitaxial material and the substrate and $\sigma_{\mathrm{interface}}$ is the free enthalpy related to the specific interface, mainly taking into account the lattice mismatch induced strain.
We may distinguish three cases (Fig. 10.5):

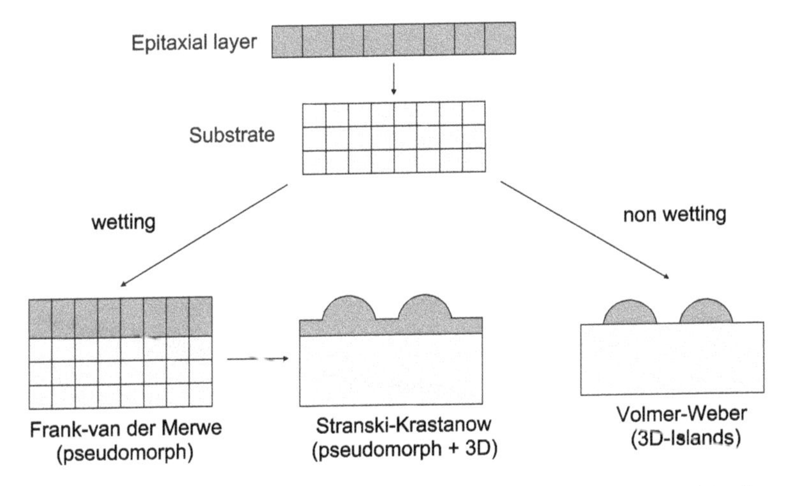

Figure 10.5 The three growth modes (schematically) as discussed in the text.

(a) $\Delta\sigma < 0$: Excellent wetting and hence two-dimensional growth can be obtained. This growth mode is called **Frank-van-der-Merwe mode** [65]. If the lattice constants of layer and substrate are different, then a pseudomorphically strained layer grows (c.f. Chapter 9).

(b) $\Delta\sigma > 0$: Due to the dominance of σ_{epi}, no wetting takes place, the layer grows three-dimensionally, i.e., many nuclei start growing without immediate coalescence (**Volmer–Weber mode**, [66]).

(c) In some cases, the growth starts under conditions where $\Delta\sigma < 0$. However, in the case of lattice mismatch, the strain generated at the interface leads to a steady growth of $\sigma_{\mathrm{interface}}$, eventually changing the sign of $\Delta\sigma$ to $\Delta\sigma > 0$ after reaching a critical thickness h_{SK}. This means: First, a strongly strained, but two-dimensional layer (called *wetting layer*) grows, on top of which after a while islands nucleate. This mode is called **Stranski–Krastanov growth mode** [67]. Hence, the island formation depends on the strain, i.e., the lattice mismatch. Also the wetting layer thickness depends on the strain, similarly as the critical thickness discussed in Section 9.3 (c.f. Fig. 9.6). However, these two

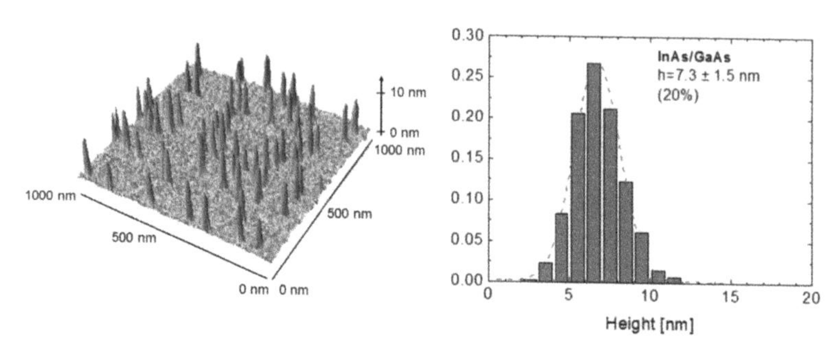

Figure 10.6 Atomic force microscopy image of InAs quantum dots on GaAs (left) and their height distribution (right, after [69]).

critical thicknesses are not identical. The island formation is not driven by defect generation.

The latter case has evolved as a very powerful method for the growth of quantum dots. Interestingly, the sizes of these self-assembled quantum dots are fairly uniform in some material systems (Fig. 10.6). The height and density of the dots are interrelated by some diffusion constant D describing the movement of the adsorbed species on the surface before it is incorporated [68]:

$$D = \frac{2kT}{h} e^{-\frac{E_d}{kT}}$$

This diffusion is balanced by the deposition rate R. Hence, the dot formation process is a function (besides the specific materials) of the temperature and of the specific lattice plane.

Best growth temperatures for quantum dots are typically about 100°C or so below "normal" growth temperatures (InAs-GaAs: $T \sim 500 - 550°\text{C}$). Therefore, MBE is more attractive as growth method, as it does not require some cracking of hydrides or so. However, excellent dots can also be achieved by MOVPE.

Such self-assembled quantum dots can be buried by another epitaxial layer (same material as "substrate") which grows two-dimensionally thus flattening the surface. Then, the dot growth can be done again, eventually leading to a stack of quantum dots. If the barrier between the dots is not too thick, then the strain field of the buried dots will define the position of the next dots. Hence, the dots are vertically aligned, and also their shape gets even more uniform.

10.3 Properties of Self-Assembled Dots

As mentioned above, the self-assembly process depends—besides the growth conditions—on the material properties. Excellent quantum dots could be obtained in the material systems:

- InAs-GaAs
- Ge-Si
- InP-GaInP
- GaSb-GaAs
- several II-VI compounds

Such dots typically have the shape of a lens or a flat pyramid with typical heights around 5–10 nm and lateral size of some 10 nm. The dot density varies between 10^8 and 10^{11} cm^{-2}. As shown in Fig. 10.6, the size of the quantum dots in these systems can be quite uniform. However, it is clear, that it fluctuates more than in lithographically defined dots. Accordingly, their optical properties reflect the variance of the dot size which directly influences the quantization. Thus, the photoluminescence signal of a dot ensemble (as measured in normal PL experiments with a spot size of at least several 10 μm) is fairly broad (Fig. 10.7). It consists of very sharp lines of the many single dots which can be made visible in single dot spectroscopy, e.g., by covering the sample surface with a mask which has nanometer sized openings.

The zero-dimensional character of these structures has been proven in many experiments. Single dots are also quite important for the investigation of quantum phenomena like quantum state entanglement or single photon light emitters.

However, quantum dots did not yet revolutionize "normal" optoelectronic devices, in particular laser diodes (as did the quantum well). The major reasons are

- their size fluctuations;

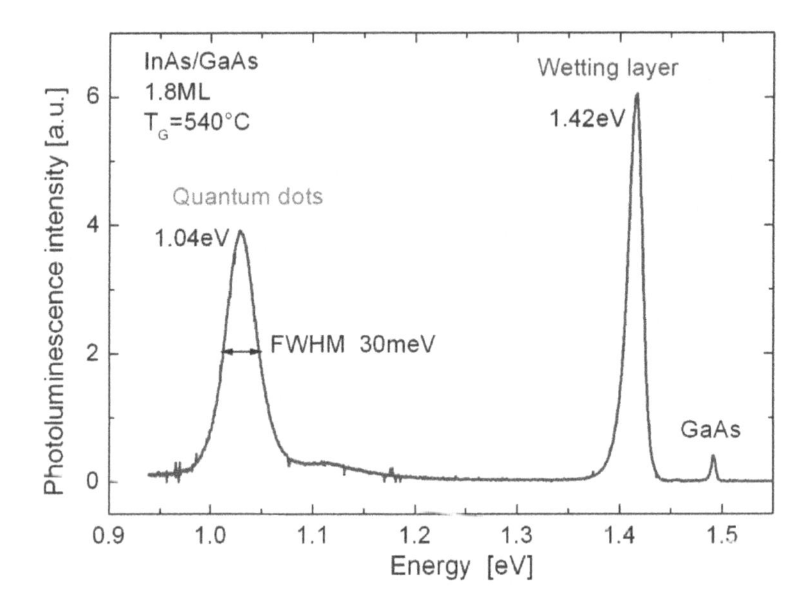

Figure 10.7 Photoluminescence signal of a sample containing self-assembled InAs dots embedded in GaAs (after [69]).

- the comparably bad control of the dot growth resulting in fairly non-reproducible device properties;
- the comparably low dot density yielding to an extremely low filling factor of active material in a laser cavity.

The latter problem can be compensated partly by stacking several dot layers (see above). Indeed, some groups have successfully realized laser diodes with very low threshold currents.

Today, it seems that quantum dot lasers are not replacing other laser diodes. They may complement them in specific areas, in particular in those cases, where they make wavelength regions accessible which cannot be accessed by quantum wells due to, e.g., lattice mismatch problems. This is in particular the case for the telecom wavelength region (1.3–1.5 μm): Such wavelengths require InP-based heterostructures (see Fig. 2.3). However, due to the much more advanced development of GaAs, lasers based on this material system would be preferred. InAs-GaAs quantum dot structures help to open up this long wavelength region for GaAs-based devices.

Problems

(1) What is the main motivation to investigate low-dimensional systems regarding optoelectronic device applications?
(2) Concerning the number of bound states in low-dimensional systems, what basic difference can be found between quantum wells and quantum dots?
(3) Determine the minimum radius for a spherical quantum dot to obtain one bound state for electrons in a GaAs conduction band quantum well embedded into $Al_{0.25}Ga_{0.75}As$ barriers.
(4) Which other (very basic) material systems have a very similar density of states as quantum dots? Explain.
(5) Explain the term "phonon bottleneck."
(6) Describe the fabrication procedure for self-assembled quantum dots as opposed to lithographically fabricated quantum dots. Write down some Pros and Cons for both.
(7) (a) What photoluminescence signal do you expect from a quantum dot?
    (b) How does the signal typically look like for a self-assembled quantum dot structure?
(8) Why could quantum dot structures be helpful for laser diodes applied in long-distance data communication?

Chapter 11

Group III Nitrides

Due to the fairly low band gap of the "conventional" compound semiconductors (arsenides, phosphides, see Fig. 2.3), only longer wavelength optoelectronic devices can be made from them:

- infrared LEDs and lasers
- red and yellow LEDs using the direct semiconductor AlGaInP for high-brightness applications
- green LEDs of fairly low efficiency using the indirect semiconductor GaP by applying nitrogen doping
- red lasers ($\lambda > 630$ nm) made from GaInP-AlGaInP

For a long time, shorter-wavelength materials were heavily wanted but not available. A lot of research went into II-VI compounds such as ZnS, ZnSe, MgS, and their ternary and quaternary alloys. However, these materials still suffer from big problems, e.g.,

- low device stability, short device life times
- strong material problems (e.g., n- and p-doping).

Shorter-wavelength optoelectronics are, on the other hand, mandatory for many applications:

- Short-wavelength lasers and LEDs, in particular for the whole visible spectral range (700–450 nm, i.e., $E_g \sim 1.8 \ldots 2.75$ eV),

Compound Semiconductors: Physics, Technology, and Device Concepts
Ferdinand Scholz

ISBN 978-981-4774-07-9 (Hardcover), 978-1-315-22931-7 (eBook)
www.panstanford.com

- lasers for high density optical data storage, as the storage density on a compact disk increases with $1/\lambda^2$),
- applications in the UV: UV light sources, detectors, etc.
- high-band gap electronics for, e.g., high-temperature applications.

When looking to the III-V compound semiconductors, then the obvious solution are the group III nitrides. Notice the trend to larger band gap when moving upward in the V-th column of the periodic system of elements:

$$\begin{aligned}\mathrm{GaSb}(E_g = 0.726\,\mathrm{eV}) &\rightarrow \mathrm{GaAs}(1.424\,\mathrm{eV})\\ &\rightarrow \mathrm{GaP}(2.26\,\mathrm{eV}) \rightarrow \mathrm{GaN}(3.39\,\mathrm{eV}). \end{aligned} \tag{11.1}$$

This is known since many decades, but also here, heavy material problems have impeded a development. These problems started to be overcome in the late 80s and early 90s of the last century, as recognized in 2014 by the Physics Nobel Prize granted to S. Nakamura, I. Akasaki, and H. Amano. Since then, the group III nitrides have become a central topic of research and development. Highly efficient LEDs emitting in the visible range of our spectrum have revolutionized many areas including novel solutions for general lighting. Moreover, GaN-based laser diodes and field effect transistors have penetrated the market. Since a while, the group III nitrides have by far surpassed GaAs as the most important semiconductor family behind Si. However, many of its basic properties are different to those of the arsenides or phosphides. All these facts justify that an extra chapter of this lecture is devoted to them.

Again, only a brief overview is given in this book. For more details, plenty of review papers and other books are available, e.g., refs. [70–72].

11.1 Basic Properties

Due to the very small ionic radius of nitrogen as compared to P, As, Sb, etc., many basic properties of the nitrides differ strongly from those of the respective other compounds.

Their most stable crystalline structure is the wurtzite modification with hexagonal symmetry, whereas most other III-V compounds crystallize in the cubic zinc blende structure, as discussed in earlier chapters.[a] Microscopically, the difference is not that big: Here and there, the atoms form tetrahedric bonds to their nearest neighbors. As mentioned in Section 1.3, this results in a dense packing of spheres. Zinc blende (ZB) and wurtzite (W) just differ in the stack sequence: ABC-ABC (ZB) or AB-AB-AB (W). Or in other words: The tetrahedron of the next neighbor has either the same orientation or is rotated by 30°. Consequently, the primitive unit cell (Wigner–Seitz cell) contains two Ga atoms and two N atoms, whereas the Wigner–Seitz cell of GaAs just contains one atom of each sort.

The most appropriate crystalline unit cell is a hexagonal prism with an in-plane lattice constant a and an out-of-plane lattice constant c (see Fig. 11.1). Please note: Hexagonal crystals have a reduced symmetry (as compared to cubic crystals). The axis of the prism (perpendicular to the basal plane) defines the so-called c-direction. This direction is more or less equivalent to the $\{111\}$-direction of the zinc blende (ZB) structure. However, only one c-direction exists, whereas cubic crystals possess four equivalent $\{111\}$-directions.

Similar to the $\{111\}_A$ and $\{111\}_B$ plane of GaAs, the hexagonal nitrides have a Ga-terminated and a N-terminated c-plane (also denoted as c-plane and -c-plane), where the respective atoms form three bonds into the crystal to the next opposite atoms and just one "dangling" bond to the open surface. Consequently, these planes have very different properties.

Of course, planes and directions in a wurtzite crystal can be indicated by their Miller's indices as discussed in Section 1.3. However, typically four "Miller's" indices are used where three indicate the respective planes perpendicular to the basal plane and the fourth indicates the basal c-plane. As only three are necessary, they are linearly dependent: The third index is just the negative sum of the first two indices.

The characteristic c-plane is then indicated as (0001) plane, whereas the c-direction is called [0001] direction. However, other

[a] The nitrides can also be forced to crystallize in zinc blende modification. However, this is only a metastable modification and has not yet achieved big technical importance.

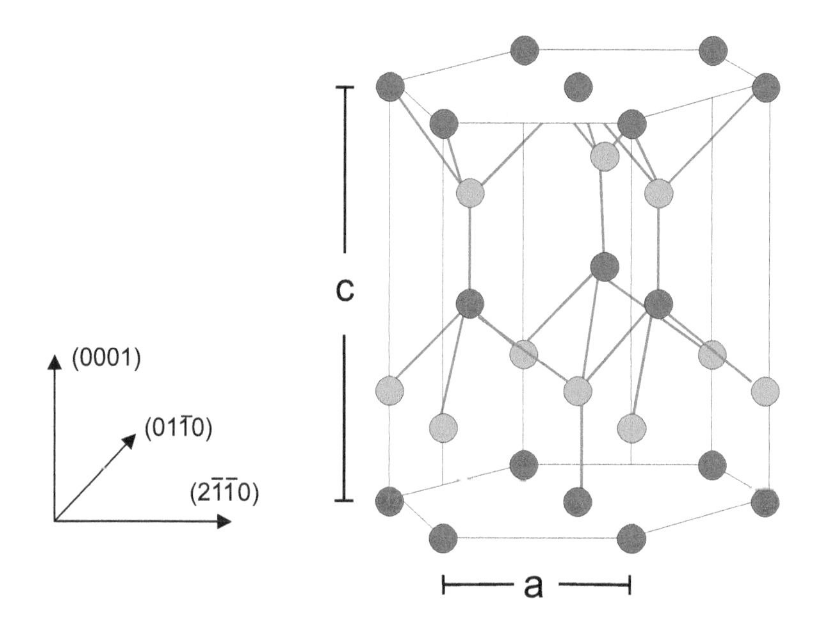

Figure 11.1 Unit cell of the wurtzite crystal.

planes and their surface normal directions may not be indicated by the same indices.

In ideal wurtzitic (W) crystals with perfect tetraedric bonds, the c and a lattice constants would be directly related by geometric laws leading to an expected c/a ratio of $\sqrt{8/3} \approx 1.633$. However, the c/a ratios in the real crystals differ from that number as a consequence of the quite strong ionic binding force between the atoms (see Table 11.1). This together with the broken symmetry leads to some spontaneous electric polarization as discussed later (Section 11.6).

The different symmetry of the hexagonal crystal also influences the band structure: As discussed for quantum wells and strain, the reduced symmetry (as compared to cubic crystals) gives rise for lifting the valence band degeneracy which we know from ZB bulk crystals. Hence, three different valence bands are formed in W crystals (c.f. Fig. 11.2). As symmetry is already broken in the bulk material, additional biaxial strain in the basal plane has much lower

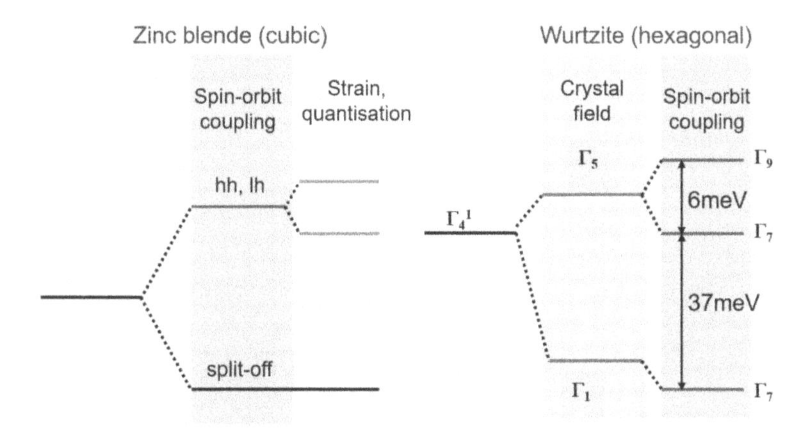

Figure 11.2 Valence band splitting due to different reasons for cubic (left) and hexagonal (right) crystals (schematically).

influence on the band structure (in terms of effective mass, etc.) as compared to cubic materials. However, it gives rise to piezoelectric effects (see later).

The three binary compounds GaN, AlN, and InN form, together with their ternaries and quaternaries, the technically relevant class of group III nitrides. Their basic properties are listed in Table 11.1.

Again as a consequence of the small ionic radius of nitrogen, the crystalline bonds have a comparably strong ionic character. This makes the nitrides chemically and thermally very stable:

- Advantage for applications in chemically harsh environments,
- Advantage for applications at elevated temperatures,
- However, device processing (in particular conventional wet etching, etc.) becomes much more difficult.

11.2 Bulk Material, Substrates

Currently, GaN is the most important member of this material class. Therefore, it is typically needed as substrate for the respective epitaxy of device structures, etc.

Table 11.1 Basic properties of the binary nitrides, compared to GaAs

Property	Symbol (unit)	InN	GaN	AlN	GaAs
Band gap at 300K	E_g (eV)	~ 0.7	3.39	6.15	1.424
Electron eff. mass	m_e^* (m_0)	~ 0.05	0.2	$\sim 0.29 - 0.45$	0.063
Heavy hole eff. mass	m_{hh}^* (m_0)	$\sim$0.5–1.65	$\sim$1		0.5
Electron mobility	μ_e (cm^2/Vs)		$\sim$700–1000		8500
Hole mobility	μ_{hh} (cm^2/Vs)		$\sim$5–20		400
Electron saturation velocity	v_s (cm/s)		2.8×10^7		2.0×10^7
Breakdown field	E_B (V/cm)		5×10^6		4×10^5
Crystal structure		W	W	W	ZB
Cubic lattice constant	a_c (nm)		0.452		0.56533
Hex. lattice constant	a (nm)	0.3548	0.3189	0.3112	
Hex. lattice constant	c (nm)	0.576	0.5185	0.4982	
c/a ratio		1.608	1.626	1.601	

As already mentioned, the synthesis of GaN bulk material is very difficult for the following reasons:

- Owing to the strong Ga–N bonds because of the large nitrogen electronegativity, GaN has a high melting point of about 2500°C.
- However, at this temperature, GaN has a very high equilibrium pressure of about 3×10^9 Pa (c.f. Table 3.1). This means at lower pressure, it sublimes easily.
- This is partly due to the nitrogen chemistry: The evaporated nitrogen forms easily N_2, which is very stable and hence remains in the gas phase.

Hence, the conventional growth of bulk crystals can only be done under extreme conditions. Many current investigations focus on various methods of growth out of solutions (c.f. principle of LPE, Section 4.1). The closer this is done near the melting point conditions, the higher the growth rate. However, even under such extreme conditions (e.g., $T = 1850$ K, $p = 1.5 \cdot 10^9$ Pa), only small pieces of about 1 cm^2 area and some 100 μm thickness have been obtained up to now.

Currently, the most promising method is hydride vapor phase epitaxy (see later), which enables the growth of fairly thick epitaxial layers (several 100 μm) on foreign substrates which then can be used as quasi-substrates.

However, the production of such quasi-bulk substrates is very expensive. Therefore, the epitaxial growth on cheap foreign substrates is very attractive and indeed turned out to work perfectly for many device concepts including highly efficient LEDs and field-effect transistors (FETs).

Such foreign substrates should have the following properties:

- same crystal symmetry as GaN
- same or nearly same lattice constant as GaN
- similar thermal expansion coefficient as GaN to keep thermally induced strain small
- chemically and thermally stable for the required epitaxial processes
- available in high crystalline quality for a fair price

The mostly used substrates are

- Sapphire (Al_2O_3):
  Hexagonal crystal structure, excellent thermal stability, high quality available.
  However: Large lattice mismatch (16%). GaN grows on sapphire rotated by 30° which minimizes the lattice mismatch.[b] As a consequence, GaN on sapphire cannot be cleaved easily, which is a problem for the formation of devices, in particular for the mirrors of laser diodes (see later).
  Another problem is the fairly large thermal mismatch leading basically to compressive strain in GaN.
  Moreover, it is an insulator, which inhibits the formation of a backside contact for optoelectronic devices.
  Finally, it has rather poor thermal conductivity making heat sinking more difficult.
- Hexagonal SiC, which seems to have more suitable properties:
  Smaller lattice mismatch (about 3.8%), no rotated growth. Hence, cleaving along the m-plane is easily possible.
  Moreover, SiC can be easily doped n-type (for optoelectronic devices with back-side contact) or for semi-insulation (for applications in FETs). Also, its thermal conductivity is very good, again an important advantage for GaN-based FETs.
  Disadvantages: High quality SiC crystals are still very expensive. The thermal expansion mismatch is quite large and leads to tensile strain in the GaN epitaxial layer promoting cracking.
- Good results have also been obtained on Si as substrate (typically on a $\{111\}$ plane in correspondence to the GaN c-plane):0
  Such wafers, heavily developed for the conventional electronic (integrated circuit) industry, are available in large size with extremely high quality for low price. Different doping is no problem.
  However, Si is plagued with a strong lattice mismatch and big difference in thermal expansion coefficient, which give rise to low GaN crystal quality, strong tensile thermal strain and

[b]Otherwise the mismatch would be 30%.

cracking even for thin epitaxial layers. Moreover, direct growth of GaN on Si leads to some direct reaction between Ga and Si ("melt-back") and the formation of SiN inhibiting further growth.
Nowadays, these problems can be overcome to a large degree by special starting procedures in the epitaxial process, e.g., an AlN buffer layer. Today, it is still under debate, whether Si may be a better substrate than sapphire for applications in optoelectronics (LEDs), whereas indeed very good electronic GaN-based devices can be grown on Si.

- GaAs is sometimes also discussed as possible substrate. Today, it is still used by a few groups as substrate for the growth of thick GaN layers by HVPE which then act as quasi-substrates. Advantage of GaAs: The thick GaN can easily be removed from GaAs (even by wet-etching, if necessary).
- There are a lot of other more exotic materials, which may have less lattice mismatch or other interesting properties, e.g., allowing the GaN growth in other than c-direction (see later). However, none of them has yet proven to come close to sapphire or Si in terms of GaN quality and applicability for devices.

Today, the main application of GaN is in high brightness blue, green and white LEDs. For these devices, nearly exclusively sapphire is used as substrate.

11.3 Hetero-Epitaxy, Defect Reduction

On these foreign substrates (discussed in the last section), group III nitride epitaxial heterostructures are grown. The main epitaxial method is metalorganic vapor phase epitaxy, applied as well in research as in industrial production, enabling the realization of all kinds of structures. Molecular beam epitaxy has a significantly lower importance, mostly just in research with a stronger focus on electronic devices, i.e., GaN-AlGaN hetero structures.

Due to the different properties of substrate and epitaxial material (as discussed in the last section), a huge density of defects is generated at the interface. From the lattice mismatch, a dislocation

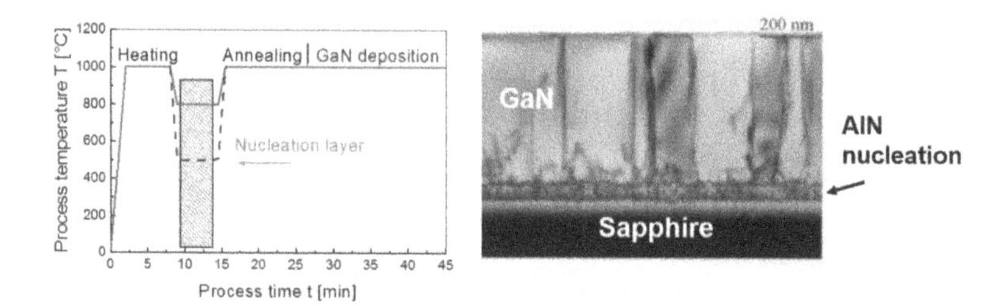

Figure 11.3 Temperature scheme of a MOVPE process to grow GaN with a low-temperature nucleation layer (left); transmission electron microscopy (TEM) image of the cross section of a GaN layer grown on sapphire with an AlN nucleation layer (right). TEM performed by P. Galtier, Thomson CSF.

density in the range of 10^{11}–10^{12} cm^{-2} can be expected. Those are typically dislocations starting at the substrate surface vertically running through the whole epitaxial structure up to the surface and hence are called "threading dislocations."

A significant break-through for the development of GaN-based structures was the introduction of specific nucleation layers, which help to reduce these defect densities dramatically (Fig. 11.3): On the foreign substrate, first a low-temperature **nucleation layer** at 500–600°C (GaN) or 600–800°C (AlN) is deposited, before the main GaN layer is grown typically at 1000°C (MOVPE). This leads to dislocation densities in the range of 10^9 – 10^8 cm^{-2}.

In order to reduce the defect density further, different approaches have been studied. One particular class of methods can be subsumed under the headline "Epitaxial Lateral Overgrowth" (ELOG), where those threading dislocations get blocked by a dielectric mask. The original method can be described as follows (Fig. 11.4 left): The (highly defective) GaN buffer layer grown in a first step is covered with SiO_2 or Si_xN_y as mask material. After lithographically opening stripes in the mask with a periodicity and width of a few micrometers, another epitaxial step leads to the growth of GaN stripes by selective epitaxy in the openings. These stripes later grow laterally over the mask eventually coalescing and closing the surface completely. This leads to potentially defect-free material over the masked areas, whereas the defects are still present in the "window regions," i.e., the stripe openings.

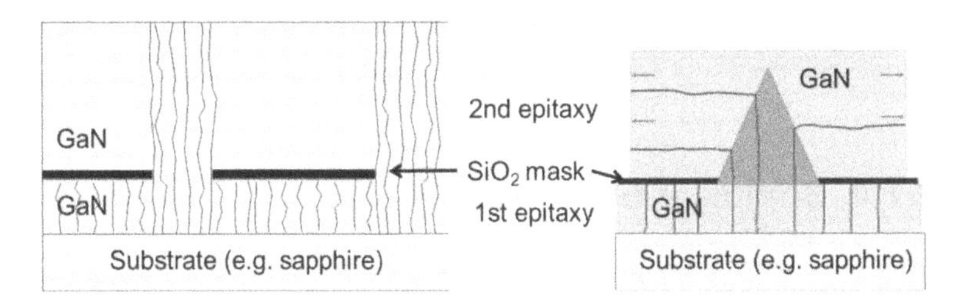

Figure 11.4 Epitaxial lateral overgrowth (left) and facet-assisted epitaxial lateral overgrowth (right, schematically).

By adjusting the growth conditions accordingly, the GaN stripes can be grown with triangular cross section. Then the process can be optimized for lateral growth leading to a lateral bending of the vertical defects which then run parallel to the sample surface. Hence, now the GaN surface is totally defect-free (**Facet-assisted epitaxial lateral growth, FACELO**; Fig. 11.4 right). With these methods and other approaches based on them, defect densities below 10^7 cm^{-2} can be obtained.

In particular for laser diodes, even lower defect densities are mandatory. Then, foreign substrates should be avoided; GaN "homo substrates" are required. As GaN bulk crystals cannot be grown by the methods described in Chapter 3 mainly due to the huge equilibrium vapor pressure of GaN at its melting point (see Table 3.1), such quasi-substrates are also mainly grown epitaxially by hydride vapor phase epitaxy (c.f. Section 4.2), where growth rates of several 10–100 μm/h can be obtained. For low defect density, ELOG or FACELO material may be used as template. Moreover, the defects get strongly reduced, the thicker the layer grows (provided that the best growth conditions are applied).

Typically, the thick GaN layer is finally separated from the starting substrate (sapphire or GaAs) to reduce any problems with thermal mismatch between foreign substrate and nitride epitaxial structure.

Another well-developed method for quasi-bulk GaN is the so-called "ammono-thermal" method, where the Ga- and GaN constituents are dissolved in supercritical ammonia. Here, crystal growth occurs at fairly low temperatures (500–600°C) and still acceptable pressures of 200–500 MPa. The interested reader may find more details about this method in the literature, e.g., [73, 74]. A nice review about GaN bulk crystal growth is given in ref. [75]

11.4 Doping of GaN

Due to the large band gap of the nitrides, also the dopants' ionization energies are larger than for phosphides and arsenides.

This is not so much a problem for **n-type doping**, which is typically achieved with Si. An ionization energy of about $E_{\mathrm{d}} \simeq$ 25 meV has been determined, which is still small enough to provide nearly complete ionization at room temperature.

The situation is more difficult for **p-type doping**. In fact, over some decades, it was unclear how GaN could be doped p-type. This fact is, besides the hetero-epitaxial problems, the main reason why the group nitrides were regarded as exotic materials without commercial value until the beginning of the 90s of the last century.

Only then, a Japanese group discovered, that Mg is a suitable p-type dopant, but an activation step is always required [76]: After the epitaxial growth, the dopant is passivated by forming a center in the crystal which additionally contains a hydrogen atom. To ionize the Mg acceptor, it is necessary to remove the hydrogen by out-diffusion at elevated temperatures (around 700°C).[c]

Even then, this acceptor has a very high ionization energy around 160–200 meV (c.f. Table 5.1). Hence, only a small fraction of acceptor atoms is ionized at room temperature (remember: $p \sim e^{-\frac{E_a}{2kT}}$). Typical best values achieved today:

- Hole concentration $p \sim$7–9 $\times 10^{17}$ cm^{-3} for Mg concentrations of 3–5 $\times 10^{19}$ cm^{-3}
- Mobility $\mu \sim$ 8–10 cm^2/Vs
- Specific conductivity $\sigma \sim$ 1.3 $(\Omega\mathrm{cm})^{-1}$

Higher Mg concentrations do not result in higher hole concentrations. Instead, the hole concentration decreases again due to self-compensation phenomena.

[c]Similar effects are also known from acceptors in other III-V compounds. The activation process is then typically integrated in the MOVPE process.

Moreover, the formation of good ohmic contacts on p-GaN is hindered by the respective work functions of the metals. Thus, the p-contact resistivity is still a significant limitation in many GaN devices.

11.5 Ternary Alloys

The epitaxial growth of $Al_xGa_{1-x}N$ is not a big problem for not too high x-values. However, a major difference to $Al_xGa_{1-x}As$ is the significant lattice mismatch between the binaries GaN and AlN leading to large strain and even cracks in heterostructures, if the thickness is too large.

AlGaN with higher Al content is in particular required for ultraviolet LEDs and detectors. Owing to the stronger reactivity and lower surface diffusion mobility of Al, higher growth temperatures are required. On the other hand, TMAl is sensitive to parasitic gas phase reactions with NH_3. All these details make the growth of Al-rich AlGaN heterostructures challenging. Currently, research toward efficient LEDs emitting at wavelengths below 300 nm is a hot topic. LEDs with emission wavelengths down to about 250 nm have been reported.

For devices working in the visible range of the optical spectrum, $Ga_{1-x}In_xN$ is required. Here, the lattice mismatch between the binary end-members is as large as 11 % leading to significant strain in GaInN quantum wells embedded into GaN barriers (for some consequences see Section 11.6).

However, thermodynamic calculations show, that this ternary system possesses a large miscibility gap for $\sim 0.15 < x < \sim 0.9$. This leads to increasing composition fluctuations in $Ga_{1-x}In_xN$ epitaxial layers with increasing x.

Moreover, due to the fairly low equilibrium vapor pressure of In, it is incorporated into the growing GaInN layer only at significantly reduced temperatures. Consequently, particularly when grown by MOVPE, the In composition x in the solid depends heavily on the growth temperature (Fig. 11.5), much different as compared to other III-III ternaries (c.f. Fig. 4.13). This leads to the following consequences:

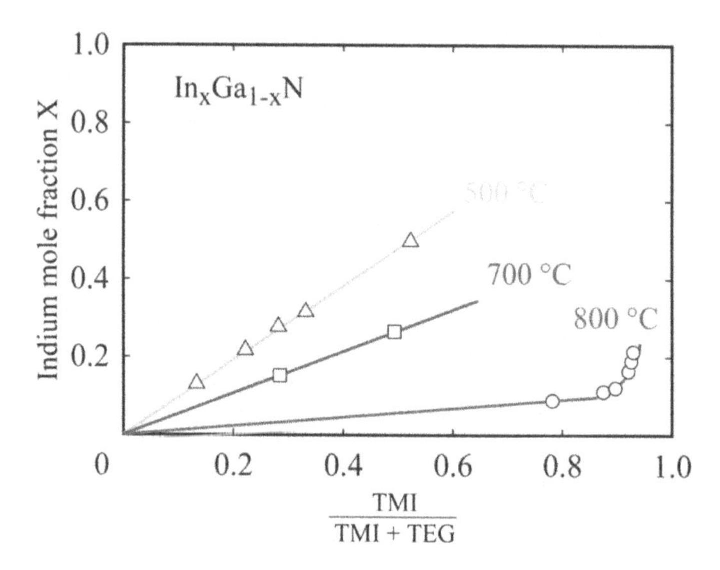

Figure 11.5 Solid composition of $Ga_{1-x}In_xN$ versus molar gas phase composition in MOVPE (from [77], with permission of Springer).

- Significant In incorporation can only be achieved for temperatures below 800°C, while the typical growth temperature of GaN is 1000°C(MOVPE).
- Even for not too large In content ($x \sim 0.1$–0.15) such low temperatures are required which is far away from the best temperature required for GaN, limiting the quality of GaInN further.

Researchers also found out that the In incorporation efficiency drops dramatically in the presence of H_2 in the MOVPE process. Hence, GaInN layers should be grown under N_2 atmosphere.

11.6 Polarization and Piezoelectricity

All III-V compound semiconductor crystals do not have inversion symmetry. Hence, they typically have piezoelectric properties, as discussed below.

This is particularly true for GaN and its alloys. Owing to the strong ionic character of the bonds in the nitrides and their specific hexagonal crystal structure, these materials possess additionally some strong spontaneous polarization characteristics.

When looking along the c-direction (which is the main growth direction for the nitrides), pure Ga and pure N sublayers can be identified in the crystal. Hence, two opposite faces (N-face and Ga-face) can be distinguished on either surface, which in principle give rise to surface charges of opposite sign. The wurtzite crystal structure is the structure with highest symmetry compatible with such a spontaneous polarization $\vec{P}_{\text{spontan}}$.

The spontaneous polarization depends on the details of the lattice parameters and hence differs for the different nitrides. One important parameter is the ratio of the c- and a-lattice constants. For perfect tetrahedral bonds, as realized in cubic zinc blende crystals, a ratio of $c/a = \sqrt{8/3} \approx 1.633$ can be determined. However, in the nitrides, this ratio deviates slightly from this value (c.f. Table 11.1).

In normal bulk samples, this intrinsic dipole moment is neutralized by "free" electric charge that builds up on the surface by internal conduction or from the ambient atmosphere. This spontaneous polarity may become visible as some kind of electric field if any "perturbation" of the material is performed, e.g., heating or cooling. Hence, such polar materials are often also named **pyroelectric materials**.

In heterostructures, the spontaneous polarizations are not necessarily compensated, but add accordingly at the hetero-interfaces, giving raise to fairly large internal electric fields (see below).

If strain is applied to such crystals in some specific directions, such charge asymmetry and hence the polarization may be altered. Hence, a strain-dependent electric field is generated, or in other words, these materials are **piezoelectric**, as shortly mentioned above.

This can be best understood for uniaxial strain applied in c-direction (c.f. Fig. 11.1): In such a strain situation the 'vertical' bonds change differently than the 'inclined' bonds, leading to a relative shift of the positive and negative ion charges in opposite directions.

Please notice: As the GaAs <111> direction corresponds to the c-direction of GaN, the same phenomenon takes place in GaAs. Hence, also GaAs is piezoelectric, but it does not have any spontaneous polarization, as it has perfect tetrahedral bonds.

For not too large strain values, the resulting electric polarization $\vec{P}_{\mathrm{piezo}}$ is linearly related to the strain ε_{jk} by the tensor of the piezoelectric coefficients e_{ijk}:

$$P_{\mathrm{piezo},i} = \sum_{j,k} e_{ijk} \cdot \varepsilon_{jk}$$

Similar as for the stress-strain relations, also here Voigt's notation can be used to simplify the formula:

$$P_{\mathrm{piezo},i} = \sum_{j} e_{ij} \cdot \varepsilon_{j} \tag{11.2}$$

For the highly symmetric wurtzite crystal, only three piezoelectric coefficients remain: $e_{31} = e_{32}$, e_{33}, and $e_{15} = e_{24}$. These coefficients are about 10 times larger for GaN than those of GaAs, and AlN possesses the largest piezoelectricity of all tetrahedrally bonded semiconductors.

As discussed in Chapter 9, the main "intrinsic" reason for strain in such semiconductor epitaxial structures is the lattice mismatch induced biaxial strain. This is also the case here: Due to the large mismatch between the different binaries, huge piezoelectric fields are present even for fairly small amounts of the "third" component, e.g.,

$$E_z = \frac{P_{\mathrm{piezo},z}}{\varepsilon_0 \varepsilon_{\mathrm{r}}} = \frac{1}{\varepsilon_0 \varepsilon_{\mathrm{r}}} \left(2e_{31}\varepsilon_{||} + e_{33}\varepsilon_{\perp} \right) \sim x \cdot 10 \frac{\mathrm{MV}}{\mathrm{cm}}$$

for x in $\mathrm{Ga}_{1-x}\mathrm{In}_x\mathrm{N}$ pseudomorphically (biaxially) strained with respect to GaN. As mentioned above, these effects are stronger than in GaAs (in the corresponding <111> direction). Moreover, they are opposite in sign. Microscopically, the reasons are somewhat different:

- In GaAs, the strain leads mainly to a polarization of the ions forming the crystal, whereas
- in GaN, the displacement of the ions is the main reason for the piezoelectric effect.

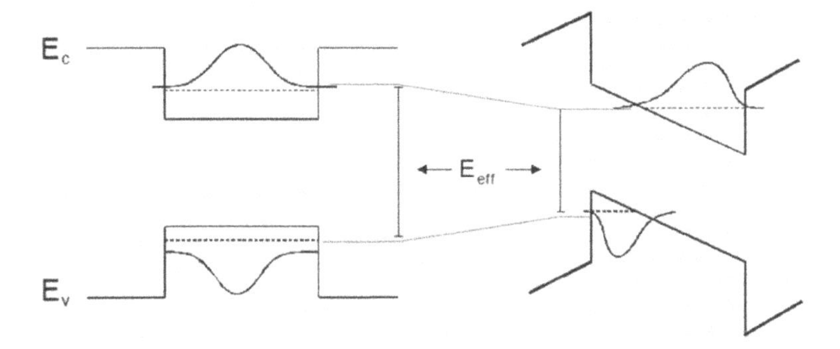

Figure 11.6 Quantum-confined Stark effect: Quantum well without electric field F (left) and with applied electric field (right; schematically). Assuming a compressively strained GaInN quantum well, the N face is on the left (typically the substrate side), the Ga face on the right (the surface side of the epitaxial structure).

Such piezoelectric fields are in some sense omnipresent in any nitride heterostructure. This is in particular true in GaInN quantum wells embedded in GaN. Owing to the larger lattice constant of InN as compared to GaN, such quantum wells are compressively strained. The piezoelectricity gives rise to an electric field pointing from the substrate side (N face) to the surface side (Ga face), which leads to a corresponding inclination of the conduction and valence band edges in the quantum well, respectively (see Fig. 11.6). This directly leads to a redistribution of the electrons and holes, driving the former to the upper surface while the latter are confined at the lower surface. Hence a separation of the carriers takes place decreasing the overlap of the respective wave functions and hence the recombination probability. Moreover, the effective band gap shrinks. This effect is named **quantum-confined Stark effect**. It may be also generated by an externally applied electric field. Then it builds the basic function of an electro-absorption modulator: By changing the effective band gap by an applied voltage, the absorption at a given wavelength can be modulated.

In nitride quantum wells, this intrinsic effect leads to a decrease of the recombination probability in quantum wells with increasing thickness or strain (i.e., In content). This is discussed as one of the reasons, why longer wavelength nitride lasers and LEDs get less and less efficient.[d] It triggered strong research efforts to synthesize GaN-based heterostructures with lower piezoelectric polarization, mainly by turning the main growth direction out of the conventional c-direction.

Another effect of the spontaneous and piezoelectric polarization is the accumulation of carriers at interfaces, e.g., the GaN–AlGaN interface (electrons, if AlGaN is grown on GaN). This means: A two-dimensional carrier gas can be generated in the respective triangular quantum well at the interface (c.f. the situation at the GaAs–AlGaAs interface, Fig. 2.4) without any doping! This effect is used in high electron mobility GaN field-effect transistors (see Chapter 13).

11.7 Diluted Nitrides

Another interesting material combination is the mixture of GaAs and GaN. Potentially, this would make accessible the band gap region of the visible spectral range (see Fig. 2.3). However, also here a large miscibility gap exists. Only small amounts of nitrogen (less than about 5%) can be incorporated into GaAs with reasonable quality (vice versa: arsenic in GaN).

Indeed, mixing even small amounts of N into GaAs, if successfully done, has two interesting consequences:

- N decreases the lattice constant of GaAs substantially. This may be used to compensate the increase of the lattice constant when alloying GaAs with In in order to achieve a lower band gap material.
- The band gap of the alloy $GaAs_{1-y}N_y$ is governed by a huge bowing parameter (c.f. Eq. 2.2) of about 20–30 eV leading to a decrease of the band gap of GaAsN starting at GaAs for increasing N content. Hence, the simultaneous incorporation of In and N leads to a potentially unstrained, but longer wavelength

[d] This generally observed problem is often called "green gap problem," (see Section 12.1.2).

quantum well in GaAs. This material is heavily investigated as a possible active material for long wavelength laser diodes for telecommunication applications, where wavelengths of 1.3–1.55 μm are needed according to the dispersion and absorption minimum of the glass fiber, on GaAs substrate. Hence, this material competes with the InAs-GaAs quantum dot structures discussed in Chapter 10.

Problems

(1) What makes the material class of group-III nitrides so special that it is discussed in an extra chapter of this lecture? Discuss briefly

- differences from other compound semiconductors;
- specific advantages;
- disadvantages and problems.

(2) Group-III nitrides can crystallize in a wurtzitic lattice and a cubic lattice. Try to determine the cubic lattice constant of AlN from the lattice constants of the wurtzitic lattice.

(3) In GaN-based laser diodes, strain does not result in those significant improvements as in GaAs-based laser diodes. Explain why this is so.

(4) Try to estimate how many dislocations can be expected if GaN is grown on a SiC substrate.

(5) Explain some methods how to decrease the defect density of GaN grown on a foreign substrate like sapphire.

(6) (a) What makes the p-type doping of GaN difficult?
    (b) How about p-type doping of AlGaN?

(7) Which problems must be considered when growing GaInN with a considerable In content?

(8) (a) Explain: What is "spontaneous polarization," what is "piezo-electric polarization"?
    (b) Which of the following materials show spontaneous polarization, which are piezo-electric: GaAs, AlAs, InP, GaN, InN? Explain.

(c) Which physical effect is called “Stark effect”?
(d) What is the quantum confined Stark effect?

(9) (a) What are “diluted nitrides”?
(b) Which features make diluted nitrides interesting for applications?

Chapter 12

Optoelectronic Devices

Again, this chapter can only give a brief overview over this field. The interested reader can find more about this topic in many textbooks, e.g., [4, 78, 79].

Owing to the direct band structure of most compound semiconductors, their main importance is given by applications in optoelectronics, i.e., their interaction with light:

- transformation of electrical energy into light or
- transformation of light into electrical energy

Moreover, current research also works on direct manipulation of light by light with the aid of some material. However, this topic is not covered in this book.

In this chapter, we will briefly discuss the following device concepts:

- light emitters: LEDs and Laser diodes
- light detectors, solar cells
- light modulators

12.1 Light-Emitting Diode

Find more about this topic in [80].

Compound Semiconductors: Physics, Technology, and Device Concepts
Ferdinand Scholz

ISBN 978-981-4774-07-9 (Hardcover), 978-1-315-22931-7 (eBook)
www.panstanford.com

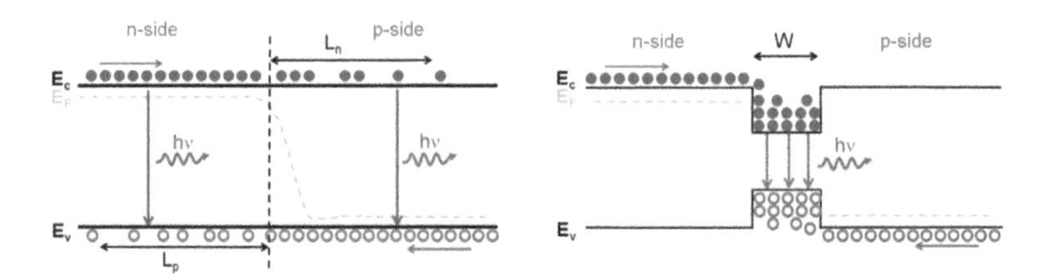

Figure 12.1 Recombination of minority carriers near a homo-epitaxial pn junction under forward bias (left). The minority carriers diffuse into the other side as governed by their diffusion lengths L_n and L_p. In a double heterostructure (right), the carriers are confined in a thin layer with width W by the potential barriers.

12.1.1 Optical Transitions

The basic way of functioning of a light-emitting diode (LED) can be explained simply: In a direct semiconductor, the electron-hole recombination which occurs in a forward biased diode (Fig. 12.1, left) at or near the pn junction, after the carriers have traversed the pn junction, is dominated by **radiative processes**. Therefore, even such a simple pn diode emits light with a photon energy close to the band gap of the respective material.

The extra carriers are supplied by an electrical current. Here, we consider the current density by normalizing our problem to a given area of the pn junction. Under steady-state conditions, the extra carrier density Δn supplied by the current density j into the region where the recombination takes place is in equilibrium with the recombination rate:

$$\frac{j}{q} = \frac{dn}{dt} \cdot d,$$

where d describes the length (perpendicular to the junction area) where both types of carriers are supplied (d times the area is the volume, where we can find the excited carriers).

Remember: The total recombination rate is given by (c.f. Eq. 6.9)

$$\frac{dn}{dt} = -R(n) = -[A \cdot \Delta n + B \cdot \Delta n^2 + C \cdot \Delta n^3] \approx \frac{\Delta n}{\tau_{\text{total}}}.$$

Hence, the excess carrier density is given by

$$\Delta n = \frac{j \cdot \tau_{\text{total}}}{q \cdot d}. \tag{12.1}$$

However, as discussed in Chapter 6, only the term $B \cdot \Delta n^2$ is responsible for light emission. For a light-emitting device, the material properties should be optimized so that radiative recombination dominates.

Consequently, for a given current density j, the non-equilibrium carrier density Δn is inversely proportional to the thickness of the active area d. On the other hand, the radiative recombination rate increases with Δn.

Hence, d should be minimized to get largest values of the recombination probability.

In a "bulk" pn **homo junction**, d is given by the diffusion lengths of the minority carriers on either side when diffusing to the other side, i.e., by intrinsic material properties: $d \approx L_n + L_p$ (Fig. 12.1, left). Typical values are in the range of 1 to 10 μm. In a **heterojunction**, d can be dramatically reduced by limiting the thickness W of the central layer with lower band gap. Hence, the recombination probability can be significantly increased by confining the carriers in such a thin layer, e.g., a quantum well (Fig. 12.1, right).

Once the (extra) carrier concentrations n and p are determined, the quasi Fermi levels can be evaluated (see Section 6.2), which then can be taken to determine the spontaneous emission rate (c.f. Eq. 6.3)

$$R_{\text{spont}}(h\nu) = \frac{1}{\tau_{\text{rad}}} \rho_{\text{j}}(h\nu) \cdot f_c(h\nu) \cdot [1 - f_v(h\nu)],$$

where $\rho_{\text{j}}(h\nu)$ describes the joint density of states and the quasi Fermi levels E_{Fn} and E_{Fp} are put into the respective Fermi distributions. Here, we assume τ_{rad} not to depend on energy.

In the Boltzmann approximation case (i.e., when the quasi Fermi levels are far from the band edges), this can be simplified yielding

$$R_{\text{spont}}(h\nu) = k_{\text{spont}}(h\nu - E_g)^{1/2} e^{-\frac{h\nu - E_g}{kT}} = \frac{(2m_r)^{3/2}}{\pi \hbar^2 \tau_{\text{rad}}} (h\nu - E_g)^{1/2} e^{\frac{\Delta E_F - h\nu}{kT}},$$

from which the linewidth of the spontaneous emission can be deduced to be about

$$\Delta\lambda \sim 1.45 \lambda_{\text{peak}}^2 \cdot kT \qquad (12.2)$$

12.1.2 Internal and External Efficiencies

The flux of photons (number of generated photons per area and unit of time) is given by

$$\Phi = R_{\text{spont}} \cdot d = d \cdot \int_0^{\infty} R_{\text{spont}}(h\nu)d(h\nu) = \frac{\Delta n}{\tau_{\text{rad}}} \cdot d, \qquad (12.3)$$

which, according to the ABC-model, is also described by the radiative term in Eq. 6.9:

$$\Phi = B \cdot (\Delta n)^2. \qquad (12.4)$$

Remember: The flux of injected carriers $j_{n,\text{injected}}$ (or e-h pairs) into the active area is given by the current density j (c.f. Eq. 12.1):

$$j_{n,\text{injected}} = \frac{j}{q} - \frac{\Delta n}{\tau_{\text{total}}} \cdot d. \qquad (12.5)$$

Hence, only the fraction

$$\eta_{\text{IQE}} = \frac{\tau_{\text{total}}}{\tau_{\text{rad}}} = \frac{1}{1 + \frac{\tau_{\text{rad}}}{\tau_{\text{nonrad}}}} \qquad (12.6)$$

of injected carriers is "transformed" into photons. That is why η_{IQE} is called the **internal quantum efficiency**, which also can be written as

$$\eta_{\text{IQE}} = \frac{R_{\text{rad}}}{R_{\text{total}}} = \frac{B \cdot n^2}{A \cdot n + B \cdot n^2 + C \cdot n^3} \qquad (12.7)$$

according to the ABC-model.

As the last term of Eq. 12.1 ($C \cdot n^3$) describes non-radiative losses, the consequence of this equation is that the efficiency of a LED decreases for large carrier concentrations, i.e., large current densities (c.f. Fig. 6.2). This problem is of particular importance in GaN-based high-power LEDs, where it is called "droop."

However, even a fraction of the generated photons cannot leave the semiconductor crystal finally. In particular those, which hit the crystal surface in an angle larger than the angle of total reflection, will be reflected back into the crystal. They may be lost by re-absorption before they get another chance to escape.

The angle of total reflectance α_{tot} is given by the ratio of the refractive indices $\bar{n}_i$ of the two respective materials:

$$\sin \alpha_{\text{tot}} = \frac{\bar{n}_2}{\bar{n}_1}$$

The refractive index of GaAs is about $\bar{n}_1 \sim 3.5$. Hence if the outer material is air ($\bar{n}_2 = 1$), then $\alpha_{\text{tot}} \sim 17°$. This means that only 4% of the photons hitting the surface under a random angle can leave the crystal, i.e., the **extraction quantum efficiency** η_{extr} is just 4%.

This low extraction efficiency can be increased by several means, e.g.,

- photon recycling, i.e., after re-absorption of a generated photon, the excited carriers recombine again radiatively;
- roughening of the surface, hence increasing the chance for leaving the material at angles below α_{tot};
- specially shaped chips (e.g., trapezoidal shape, lens, etc.);
- embedding the chip into some material with intermediate refractive index (transparent epoxy).

Some LED material systems suffer from absorbing substrates, e.g., visible light-emitting devices made of AlGaInP grown on GaAs substrates. Then, the substrate may be removed to increase the LED performance.

Another loss mechanism is given by the absorption of light by the electrical contacts. Typically, the LED current and thus light generation is strongest below the (p-)contact, which, however, acts as a mask for the generated light. Way out:

- specially shaped contacts (grids, etc.);
- transparent contacts (indium tin oxide, ITO; extremely thin metals);
- increased current spreading by thick transparent semiconductor layers (e.g., 20–40 μm GaP on AlGaInP structures);
- contact acting as a mirror, light emission on the other side.

Regarding the semiconductor material properties, the quantum efficiency is typically closely related to the defect density (e.g., dislocations). In the phosphide and arsenide material system, the quantum efficiency drops substantially for line defect densities above about 10^4 cm^{-2}. This is quite different for GaN based LEDs which show high quantum efficiencies even for dislocation densities above 10^8 cm^{-2}.

In some wavelength regions, high quality materials are not available. This is in particular true for the visible green spectral range: The highest band gap of the phosphides is about 2.4 eV (c.f. Fig. 2.3), however, the high-band gap phosphides GaP and AlP have both indirect band structure. In consequence, carrier confinement becomes more and more difficult, the shorter the emission wavelength. Hence, the quantum efficiency decreases. Additionally, high Al content materials (AlGaInP) are needed for large band gaps. Typically, the material quality decreases with increasing Al content owing to the strong reactivity of Al with many impurities, e.g., oxygen.

On the other hand, coming down in band gap energy from GaN, GaInN is needed as active material with large In content for emission in the green spectral range (c.f. Fig. 2.3). As discussed in Section 11.5, such layers suffer from the miscibility gap of GaN and InN and the increasing internal piezoelectric fields.

Both facts lead to the situation, that the quantum efficiency of LEDs in the green spectral region is much lower than that in the blue or red range (the so-called "green gap").

An additional loss mechanism may even take place outside of the active area: Not all carriers supplied by the external current really arrive in the central "active" region. This loss mechanism was not explicitly considered in Eq. 12.5. It is described by the terminus **injection efficiency** $\eta_{\text{inj}} < 1$.

Now, we can put all together: The total quantum efficiency is composed by all of these effects:

$$\eta_{\text{total}} = \frac{\text{number of emitted photons}}{\text{number of injected carriers}} = \eta_{\text{IQE}} \cdot \eta_{\text{extr}} \cdot \eta_{\text{inj}} \qquad (12.8)$$

12.1.3 Luminous Efficacy

Up to now, we have discussed the brightness of LEDs in terms of emitted photons, which can be easily calculated into emitted light power:

$$P_{\text{light}} = \int \Phi(\nu) \cdot h\nu d\nu,$$

where $\Phi(\nu)$ describes the number of emitted photons of a given frequency (i.e., wavelength) per time unit.

From this number, the so-called **wall-plug-efficiency** η_{WPE} can be calculated:

$$\eta_{\text{WPE}} = \frac{P_{\text{light}}}{P_{\text{electric}}}, \tag{12.9}$$

where P_{electric} describes the electric power consumed by the LED ($P = V \cdot I$).

This wall plug efficiency contains, beside the total quantum efficiency, also losses which may be a result of the series resistance of the device, i.e., a part of the electrical power is just transformed into Joule's heat.

However, at least for LEDs emitting in the visible part of the spectrum, another property is quite important, which tells us how bright we see the LED. This means the brightness of the LED is measured in terms of the sensitivity of the human eye, and the respective unit is the **lumen**, which describes the light or radiant flux. It is defined as follows:

1 lumen is the luminous flux of light through a solid angle of 1 steradian which is generated by a point source with 1 candela intensity.

The candela (Cd) is the basic S.I. unit of light radiation: 1 Cd = monochromatic radiation of 5.40×10^{14} Hz with a radiant intensity of $\frac{1}{683} \frac{\text{W}}{\text{srad}}$.

This definition takes into account, that the human eye is most sensitive at the green wavelength of about 555 nm ($\cong 5.40 \times 10^{14}$ Hz).

Hence, the luminous flux can be obtained as follows:

$$\Phi(\text{lumen}) = 683 \frac{\text{lumen}}{\text{W}} \int_{\lambda} V(\lambda) \cdot P(\lambda) d\lambda, \tag{12.10}$$

where $V(\lambda)$ describes the eye sensitivity (see Fig. 12.2) and $P(\lambda)$ describes the spectral power density of the emitted light in Watt/nm.

Now, the efficiency of a light source can be described in units of lumens per Watt (lm/W), which then gets the new name "luminous efficacy." The maximum obtainable efficacy, hence, depends on the wavelength: It is 683 lm/W for a monochromatic green perfect light source (555 nm) and about 250–300 lm/W for a white light source (depending on the detailed spectrum, see discussion in [82]).

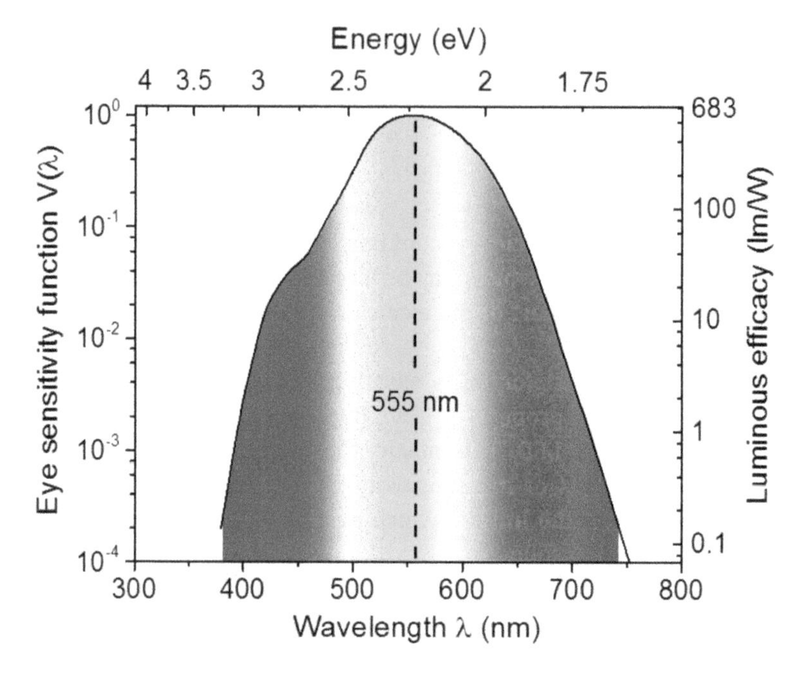

Figure 12.2 Sensitivity of the human eye (at daylight) versus wavelength (CIE 1978 $V(\lambda)$ function, after data from [81]). The right scale represents the luminous efficacy following from the definition of the SI unit "Lumen."

12.1.4 LED Colors, White LEDs

Due to the wide range of choice regarding compound semiconductors with different band gap (see Fig. 2.3), LEDs emitting over a huge range of wavelengths are now available, starting in the infrared region (GaInAs-InP, GaAs, etc.) over the visible spectrum (red and yellow: GaInP-AlGaInP; green and blue: GaInN-GaN) up to very short wavelengths in the ultraviolet region of 300 nm and below (GaN-AlGaN). Owing to their basic principle, LEDs emit monochromatic light which is related to the (effective) band gap of the respective semiconductor (with a spectral width described in Eq. 12.2).

Due to their high efficiency, LEDs are attractive light sources not only for special applications, but also for general lightning. However, here white light is required, preferably with a spectrum close to that of the sun.

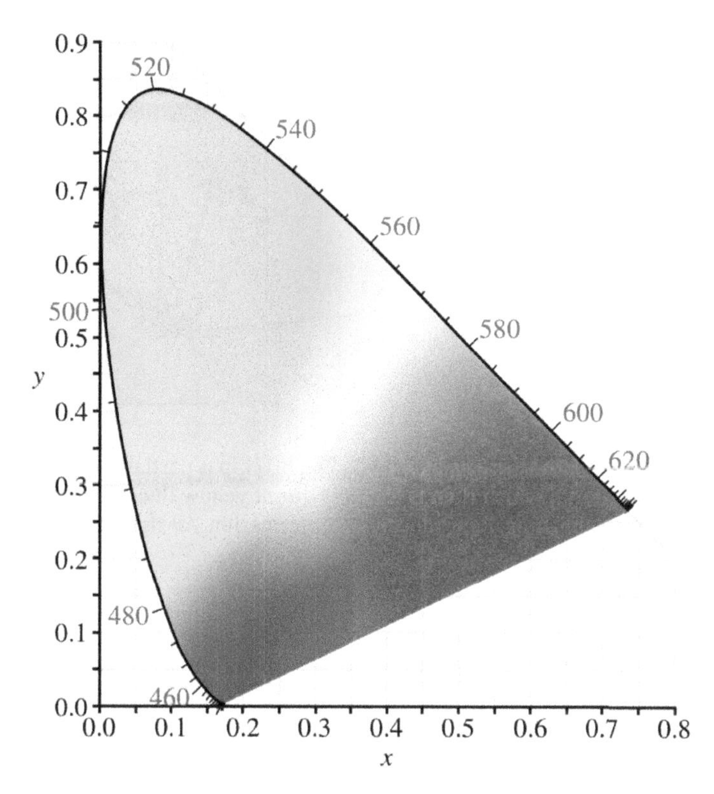

Figure 12.3 Chromaticity chart established by the International Commission on Illumination (CIE) in 1931. Every color is assigned to a two-dimensional number (x, y). The wavelengths (in nm) of the pure colors are indicated along the circumference of the figure.

One chance is the combination of three LEDs emitting the three basic colors red, green, and blue (RGB). As realized in the pixels of television or computer screens, any color can be mixed by these basic colors; in particular those which are within the triangle defined by the coordinates of these colors in the color diagram (Fig. 12.3). By mixing in the right relation, pure white can be obtained.

Since the availability of high brightness red, green and blue LEDs, large full-color TV screens could be realized with LEDs, which are now omnipresent in our cities. For some applications, systems of three LEDs are used where every particular color can be

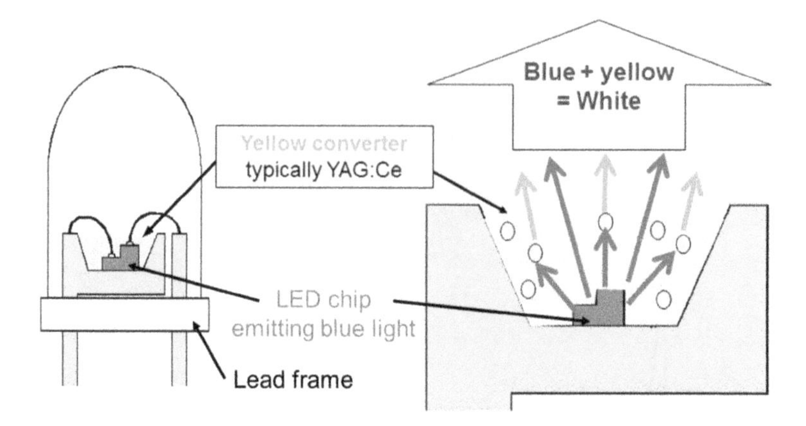

Figure 12.4 White LED: The light of a blue LED chip is partially absorbed in the cover material which then is excited to emit yellow fluorescence. After [83].

adjusted by the individual LED currents. However, this solution is—although being highly efficient—too expensive for normal lightning applications.

Some years ago, a different approach has been invented and developed since then: The blue light of a GaN LED is passed through a material which partly absorbs the blue light. The excitation energy is then re-emitted by luminescence (or fluorescence) at longer wavelengths (see Fig. 12.4). Typical fluorescence materials (or dyes) which have high luminescence conversion efficiencies are yttrium-aluminum-garnet phosphors doped with rare earth elements like cerium (Ce:YAG). The resulting light (in many cases a broad yellow emission line) is mixed with the blue light eventually giving white light. The fluorescence material is incorporated into the housing of the LED. Hence, these luminescence conversion LEDs work very similar to fluorescent lamps, where first UV photons are generated which then are transformed into visible light in a dye layer deposited on the inner surface of the glass tube.

Since the first white LEDs, the dye material has been further improved, in particular to give a "true" white light, including the red spectral part which was missing in the first attempts.

In parallel, the GaN LEDs have been improved dramatically. Best white LEDs now achieve efficiencies of more than 250 lm/W, i.e., a WPE close to or slightly above 80% [83].

12.1.5 LED Efficiencies

The further improvement of the LED efficiencies is an ongoing field of research. Currently we can observe the following:

- Longer-wavelength LEDs (red color) achieve extremely high internal quantum efficiencies of 90%. They are made of AlGaInP heterostructures. Here semiconductor material problems could be perfectly mastered. The extraction efficiency is somewhat lower. Here, the large refractive index (around 3.3) makes the outcoupling difficult. Total efficiencies of more than 100 lm/W have been achieved at 618 nm (notice the lower value of the eye sensitivity curve, Fig. 12.2, at this wavelength!).
- Short wavelength (blue) GaInN-GaN LEDs also achieve very high WPE. However, here the extraction efficiency is higher (about 80%) due to the smaller refractive index of GaN (about 2.5), whereas the quantum efficiency is still not perfect, as material issues are not yet perfectly solved.
- In between, we can identify the so-called *green gap* due to the lack of materials in this region (see Section 12.1.2).
  The latter problem still motivates further research efforts. Further improvement of the GaN LED technology by thorough optimization of GaInN based heterostructures seems to be the main road of success, whereas the search for new materials or new concepts and ideas does not show promising results.

12.2 Semiconductor Laser Diodes

Originally, the word **LASER** is an abbreviation which means: **L**ight **a**mplification by **S**timulated **E**mission of **R**adiation.

In principle, a laser is an oscillator basically very similar to an electronic oscillator used for radio frequencies. It contains all its essential features (c.f. Fig. 12.5):

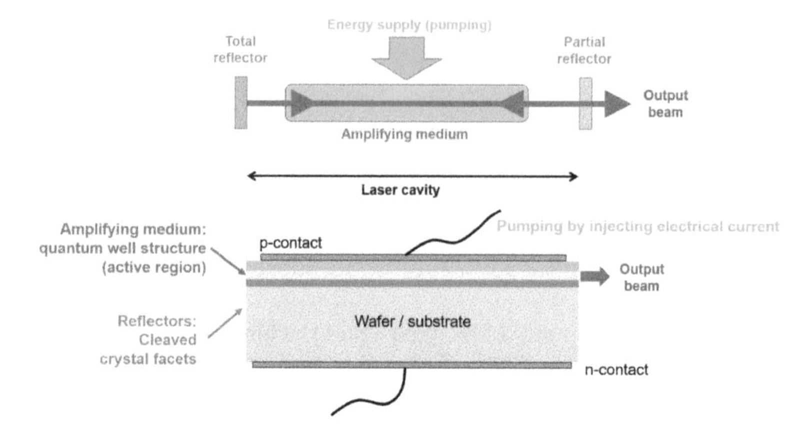

Figure 12.5 Laser principle (top) and edge-emitting semiconductor laser diode (bottom).

Feature	Laser	Electronic oscillator
Amplification mechanism	Light amplification in semiconductor	Transistor
Resonator	Front and end mirror	Resonant LC circuit
Feedback	Reflection of light	Feedback by induction in coupled coils
Energy source	Pump current	Power supply

12.2.1 Gain

The optical gain in a laser is provided by the process of stimulated emission. For light amplification, the stimulated emission must be stronger than the respective absorption (c.f. Eqs. 6.1 and 6.2). This is only possible for inversion of the carrier occupation, which is equivalent to the condition that the difference between the quasi Fermi levels is larger than the band gap:

$$E_{\mathrm{Fn}} - E_{\mathrm{Fp}} \geq h\nu \geq E_{\mathrm{g}} \tag{12.11}$$

This condition is the **first laser condition**. Only then, the number of new photons created by stimulated emission can be larger than the number of photons lost by fundamental absorption.

The extra obtained photons are then described by the gain coefficient g which is measured in inverse cm (see Section 6.3.3). It is directly related to the (excess) carrier concentration. At a given concentration of generated carriers $\Delta n = n_{\text{excited}} - n_{\text{equilibrium}}$, the material starts to amplify the light, i.e., gain just compensates absorption which is equivalent to

$$g - \alpha > 0, \tag{12.12}$$

where α describes the internal absorption coefficient. Hence, the material becomes transparent, and this carrier concentration is called the transparency carrier density n_{transp}.

In a good approximation, the gain is then related linearly to the carrier concentration above this transparency density:

$$g = A(n_{\text{excited}} - n_{\text{transp}}), \tag{12.13}$$

where A describes the differential gain factor, a kind of material constant depending on such things as density of states, band gap, effective mass, temperature, etc. (c.f. Eq. 10.6).

12.2.2 Resonator and Feedback

As described in Section 6.1, the photons generated by stimulated emission are completely identical to the stimulating photons. If the photon frequency is fixed by something, then the laser light emission contains only identical photons. This is typically achieved by two mirrors which form a resonator for a given photon wavelength so that the twice reflected photons are in phase with the original photons.

Moreover, these mirrors provide the necessary feedback by keeping an essential part of the generated photons in the resonator.

Of course, some laser light should leave the resonator through the mirrors forming the laser beam. This is an additional loss mechanism described by the reflectivity factor R_1 and R_2 of the two mirrors.

In order to keep the light generation process running, the light intensity I_1 travelling back and forth through the resonator (one round trip, i.e., a distance of $2L$) should not be weaker than the intensity I_0 at the beginning of the trip:

$$I_1 = I_0 e^{2(g-\alpha_i)L} \cdot R_1 R_2 \geq I_0,$$

where L describes the length of the laser resonator. This brings us to the **second laser condition**:

$$g \geq \alpha_\mathrm{i} + \frac{1}{L} \ln \left(\frac{1}{R} \right) \tag{12.14}$$

if the same reflectivity R is assumed for both mirrors for simplicity.

In a semiconductor laser, the resonator mirrors are typically realized by cleaved crystal facets. If these facets are not specially coated (for higher or lower reflection), then their reflectivity is given by the refractive index jump between the semiconductor (GaAs: $\bar{n}_\mathrm{s} \sim 3.5$) and air ($\bar{n}_\mathrm{air} = 1$). Thus the reflectivity is given by

$$R = \left(\frac{\bar{n}_\mathrm{s} - 1}{n_\mathrm{s} + 1} \right)^2 \sim 30\%.$$

For many applications, the mirror reflectivities can be adjusted by specific coatings to get more reflective or less reflective end mirrors.

12.2.3 Electrical Pumping

In a semiconductor laser diode, the extra energy necessary to keep the light amplification in a steady state is provided by electrical pumping, i.e., by injecting electrons and holes from either side of a p-n junction, where the active medium is placed just in the center.

This current supplies the extra carriers similarly as discussed above for LEDs. Starting at some transparency current j_transp, the gain is generated and starts to increase:

$$g = g_0 \left(\frac{j}{j_\mathrm{transp}} - 1 \right)$$

The transparency current is related to the transparency carrier concentration by

$$j_\mathrm{transp} = \frac{qd}{\eta_i \tau_\mathrm{rad}} n_\mathrm{transp}. \tag{12.15}$$

We see: For a small transparency current we need:

- small volume of pumped region, i.e., small thickness d of active layer (carrier confinement);
- large internal quantum efficiency η_i;
- large radiative life-time τ_rad.

By taking into account the additional losses described in Eq. 12.14, we finally can derive the condition for the threshold current j_{thr}, at which the laser diode starts to emit laser light:

$$\begin{aligned} j_{\mathrm{thr}} &= j_{\mathrm{transp}} + \frac{q \cdot d}{A \eta_{\mathrm{i}} \tau_{\mathrm{rad}}} \left(\alpha_{\mathrm{i}} + \frac{1}{L} \ln \left(\frac{1}{R} \right) \right) \\ &= j_{\mathrm{transp}} + \frac{q \cdot d}{A \tau_{\mathrm{total}}} \left(\alpha_{\mathrm{i}} + \frac{1}{L} \ln \left(\frac{1}{R} \right) \right) \end{aligned} \quad (12.16)$$

12.2.4 Optical Waveguide and Confinement

In order to get stimulated emission and hence optical gain, the light must interact with the active semiconductor material. If no special measures are taken, the light may leave the active region in all directions (only fed back in one direction by the laser mirrors). This problem may be overcome by embedding the active region into a wave guiding structure, i.e., some material with larger refractive index in which the light is confined due to total reflection at the interface to the surrounding material with lower refractive index.

Historically, the improvement of the waveguide design was a key feature in the development of laser diodes (Fig. 12.6), also reflecting the development in epitaxial methods. In the very first attempts, just p-n junctions—grown by liquid phase epitaxy—were used, where a small refractive index step was achieved just by the lower carrier concentration at the junction as the former is related to the latter according to the following equation:

$$\bar{n}_r(n,\, p) = \sqrt{\varepsilon_r} - \frac{q^2 \lambda^2}{8\pi^2 \varepsilon_0 c^2 \sqrt{\varepsilon_r}} \left(\frac{n}{m_e} - \frac{p}{m_h} \right)$$

which results in $\Delta\bar{n}/\bar{n} \sim 10^{-3}$ at a p-n junction with respect to the adjacing layers.

Later, people made use of the refractive index step at heterojunctions: A material with higher band gap E_g has a lower refractive index than a material with lower band gap. For $Al_xGa_{1-x}As$, the refractive index is given by

$$\bar{n}(x) = 3.59\sqrt{1 - 0.23x},$$

which sums up to $\Delta\bar{n}/\bar{n} \sim 3.5\%$ for a GaAs-$Al_{0.3}Ga_{0.7}As$ heterostructure.

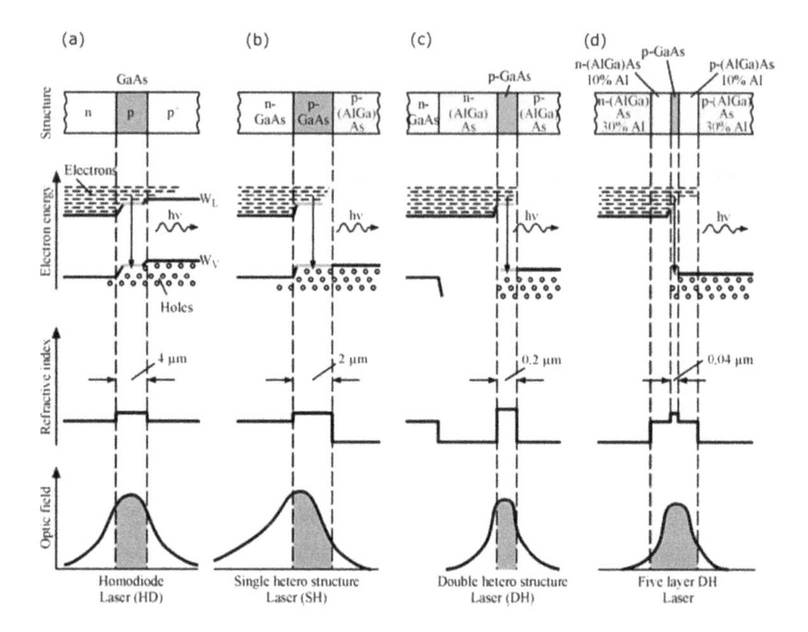

Figure 12.6 Comparison of different semiconductor laser structures: (a) homodiode laser (HD); (b) single heterostructure laser (SH); (c) double heterostructure laser (DH); (d) five layer DH laser; small refractive index steps are in the range of 0.1% to 1%, large steps at 5% [84] (from [85], with permission of Springer).

Hence, the threshold current could be improved by better optical confinement in single and later in double heterostructures. Note that such structures provide both, electrical **and** optical confinement simultaneously.

However, as discussed above, for best electrical confinement, the active semiconductor material should be very thin (quantum well) in order to increase the local carrier density. The typical scale which describes the interaction of light with matter is given by the wavelength, which typically is much larger (some hundreds of nanometers) than the quantum well thickness (few 10 nm). Hence, the waveguiding function gets lost in a quantum well structure (the light does not “see” the thin quantum well).

This problem can be reduced by providing extra layers for optical confinement. This leads to the design of a separate confinement heterostructure (SCH) (Figs. 12.6 and 12.7). Such sophisticated laser

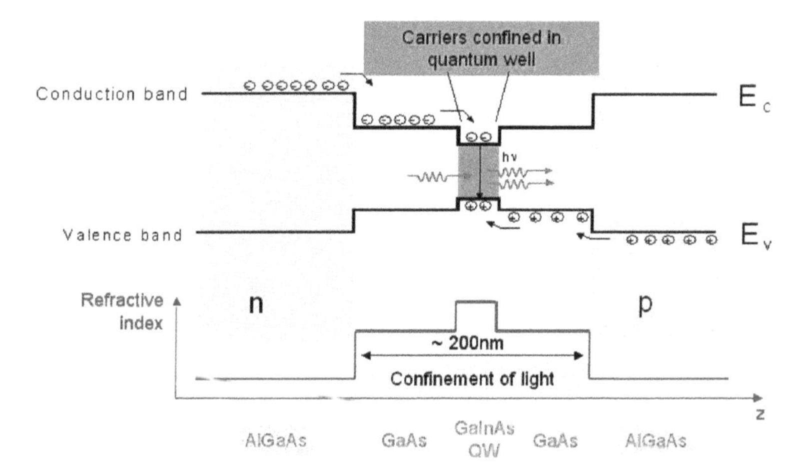

Figure 12.7 Separate confinement heterostructure (schematically): naïve band structure (top) and refractive index profile (bottom).

structures could be only grown after the development of more advanced epitaxial methods like MOVPE and MBE.

However, now the overlap of the light field and the quantum well is not perfect and becomes a design parameter for the optimization of a laser structure. It is given by the **confinement factor**

$$\Gamma = \frac{\int_{-L/2}^{L/2} |E(z)|^2 dz}{\int_{-\infty}^{\infty} |E(z)|^2 dz}, \tag{12.17}$$

which describes the overlap of the electric field of the light (photons) with the gain material in the quantum well with thickness L along the direction z (perpendicular to the quantum well plane). Typical values are in the range of a few percent for edge-emitting laser diodes.

Some design considerations:

- For small threshold current, Γ should be large to make best use of the material gain.
- However, for large material gain, the quantum well thickness ($\rightarrow$ electrical confinement) should be small.
- A thick low band gap active region (large Γ) also leads to increased losses by internal absorption α_{i}.
- Absorption losses are also often caused by absorption in the high band gap outer barriers forming the waveguide,

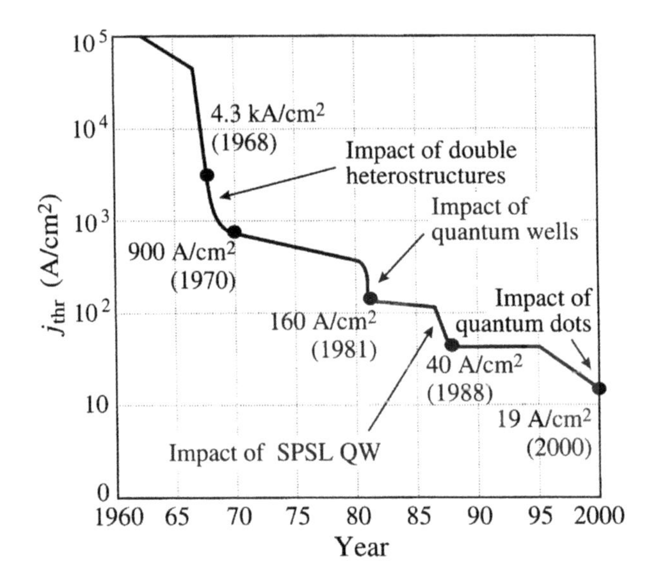

Figure 12.8 Evolution of the threshold current density of semiconductor lasers (from [86], with permission of Springer).

because these layers typically contain higher Al concentrations. A strong overlap of the light field with the barrier material can be reduced in broad waveguide structures (LOC = large optical cavity). The low α_i enables a long resonator, which is advantageous for high power lasers, because the optical gain is spread over a long distance and hence saturates not as fast as in shorter resonators.

Every step in development as described in Fig. 12.6 resulted indeed in a step-like decrease of the threshold current (Fig. 12.8). The last steps have been obtained by including strain (see Chapter 9) and self-assembled quantum dots (see Chapter 10) into the active region.

12.2.5 Temperature Behavior

Besides light absorption losses, there are more counteracting mechanisms in a laser diode, which more or less depend on temperature, e.g.,

- temperature dependence of the optical gain as a consequence of the thermal population of higher states, c.f. Fig. 10.2 (remember the improved properties over lower dimension structures);
- losses by Auger processes;
- carrier losses by thermal escape of carriers over the opposite heterojunction barrier (mainly electrons spilling over the quantum well and waveguide region into the p-doped barrier).

Such effects lead to an increase of the threshold current j_{thr} with temperature which is empirically described by the characteristic temperature T_0:

$$j_{\text{thr}}(T) = j_{\text{thr}}(T_{\text{ref}})e^{\frac{T-T_{\text{ref}}}{T_0}} \quad (12.18)$$

For standard GaAs-based lasers (emission wavelength around 900 nm), this characteristic temperature T_0 is typically in the range of 100–200 K, whereas it is much smaller for GaInAs-InP-based laser diodes for telecommunication applications (T_0 ~70–80 K), mainly due to the small conduction band offset leading to a thermal escape of the electrons out of the active quantum wells. It can be improved by using GaInAs-AlInAs-based laser diodes emitting at the same wavelength where the conduction band offset is larger.[a] Owing to the larger effective mass of the holes, they are less sensitive for such thermal escape.

12.2.6 Lateral Confinement

Similar as in the vertical direction, also in lateral direction, some kind of optimization of the overlap of the light and the gain providing region should be realized. We can distinguish two main mechanisms (c.f Fig. 12.9):

- **Gain guiding**: Light is only generated in the gain region, typically between the electrical contacts where the carrier injection takes place. Any light leaving this region laterally vanishes quickly.

---

[a] Notice that AlInAs lattice-matched to InP has a similar band gap as InP and hence can be used as a barrier material in such structures. However, the ratio between the band offsets in the conduction band (ΔE_{c}) and the valence band (ΔE_{v}; see Fig. 2.4) are about 60:40 here, whereas it is about 35:65 for GaInAs-InP heterostructures.

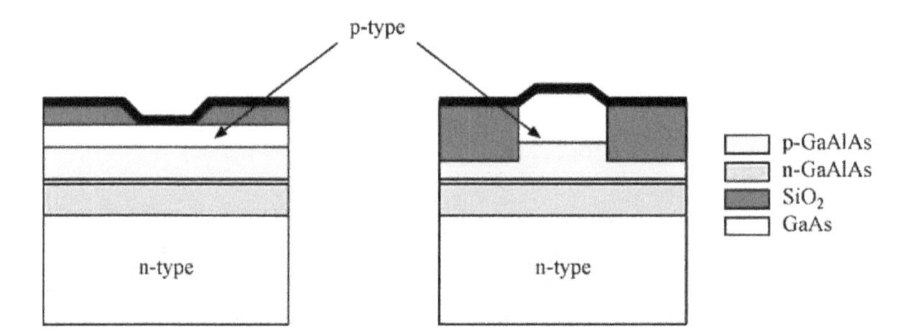

Figure 12.9 Gain-guided laser diode where the contact definition by an insulating oxide stripe defines the current flow (left) and index-guided structure where a semiconductor ridge is embedded in lower-index material (here: just SiO_2) thus defining a lateral variation of the effective refractive index (right). The black line on top represents the electrical top contact layer.

As this is only a weak guiding effect, it is mainly applied to broad area laser diodes where the gain guiding is defined just by the contact geometry (either by structurization of the metal contact or by masking the semiconductor before contact formation → oxide stripe laser). Typical width: 20–100 μm and more.

- **Index guiding**: Also in lateral direction, a wave guide structure can be realized by laterally changing the refractive index. Simplest example: Ridge laser structure: By etching a ridge, the active laser part gets (partly) surrounded by air, i.e., a zone with lower mean refractive index. In order to overcome problems with open semiconductor surfaces (which may result in non-radiative recombination losses), buried ridge structures have been developed, where the ridge is embedded into lower refractive index (i.e., higher band gap) semiconductor material in a second epitaxial step.
  Index guiding is in particular necessary for narrow stripe lasers (few μm) which provide
  - Lower threshold currents (as the pumped area is smaller)
  - Improved laser beam properties (see below).

12.2.7 Near Field, Far Field

The local intensity of the light on the outcoupling surface (typically the laser mirror facet) is called **near field**. Its structure depends on the details of the vertical and lateral structure or in other words on the waveguide properties. Similar as in microwave waveguides, the number of allowed modes depends on the waveguide structure and decreases with decreasing waveguide dimensions.

The properties of the laser beam relevant for the user are best described by the **far field**, i.e., the local light intensity falling on a plane perpendicular to the laser emission (and hence laser resonator axis) far away from the device. It is (more or less) the Fourier transformation of the near field. In a simple picture, it may be regarded as the diffraction pattern of the laser light where the outcoupling area (i.e., the cross section of the laser resonator cavity) acts as two-dimensional diffraction slit.

Hence, single (lateral) mode lasers providing best beam quality (Gaussian shape) typically can be achieved by narrow (lateral) stripes (ridge laser structures). In vertical direction, single mode is easily achieved, as the vertical waveguide structures are typically composed by adequately thin epitaxial layers.

12.2.8 Distributed Feedback Lasers

The feedback in an edge-emitting laser is typically provided by two cleaved crystal facets acting as two mirrors (see above). Hence, such a resonator is called **Fabry–Pérot resonator**.[b]

Another way of feedback can be realized by a diffraction grating integrated into the laser cavity (Fig. 12.10). This grating modulates the effective refractive index $\bar{n}_{\text{eff}}$ in the waveguiding region formed by the two materials above and below the grating structure along the laser cavity (longitudinally). Depending on the longitudinal position, the upper or lower material is thicker or thinner, thus contributing more or less to $\bar{n}_{\text{eff}}$. Now every line of the grating will reflect a part of the light travelling along the cavity, i.e., perpendicular to the stripes. The light reflected at every grid line will interfere constructively for

[b]After the two French Physicists Charles Fabry and Alfred Pérot, who developed an optical resonator made of two parallel glass plates at the end of the 19th century.

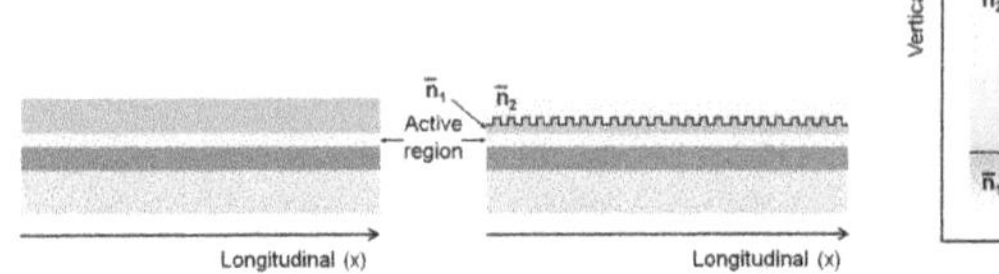

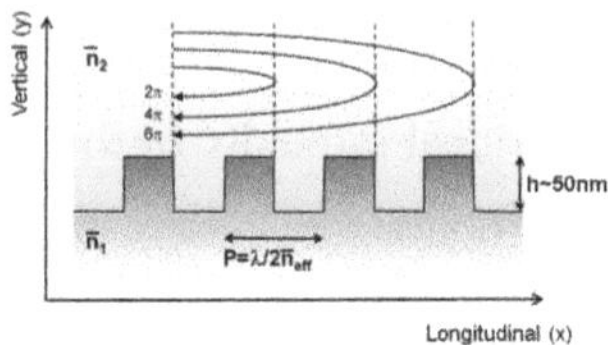

Figure 12.10 Fabry–Pérot laser with reflective facets at the ends (left) and DFB laser where the feedback is provided by a grating which modulates the refractive index near the active quantum wells (middle and right). A part of the light travelling in the laser cavity is reflected at each refractive index step and interferes constructively if the wavelength fits to the period of the grating (right).

a well-defined period P of the grating:

$$\lambda = 2 \cdot \bar{n}_{\text{eff}} \cdot \frac{P}{m}, \tag{12.19}$$

where m is the order of diffraction.

The overall reflection can be very high (depending on the number of grid lines). The same effect is achieved in vertical direction in a multi-layer structure of well-suited periodicity (see Section 12.2.9) and is effectively the same kind of interference as discussed for x-ray beams (c.f. Eq. 1.2). Hence, such a structure is also called **Bragg mirror**.

If the Bragg grating is integrated into the resonator of an edge-emitting laser diode, then this laser is called DFB laser (DFB = distributed feedback). It can be realized as follows:

- Grow laser structure epitaxially, ending in the waveguide layer slightly after the active quantum wells.
- Etch a grating into this structure (typical period: some 100 nm according to Eq. 12.19, typical depth some 50 nm; needs high-resolution lithography and in most cases dry-etching).
- Bury this grating by a second epitaxial step with another material. This leads to a longitudinal modulation of the effective refractive index which gives rise to the Bragg interference.
- Continue epitaxy by completing the laser structure (typically p-doped barrier, etc.).

If the Bragg grating is placed only at the ends of the resonator, then the laser is called DBR laser (DBR = distributed Bragg reflection).

Different possibilities for providing the feedback:

- Index coupling (as just described). Here, two stable modes develop. Some additional (symmetry breaking) feature is needed to favor one of them.
- Gain coupling: If the grating is etched through the active quantum well, then the optical gain gets modulated along the cavity. Advantage: Always only one stable laser mode. Disadvantage: Fabrication is more critical.
- Coupling by modulating the absorption. A simple fabrication method provides this kind of coupling:
  - Grow the complete laser structure epitaxially in one step.
  - Etch a (narrow) ridge.
  - Fabricate a metallic diffraction grid directly besides the laser ridge (by lithography and etching).

  The light travelling along the ridge will partly leak into the metallic grid and hence gets periodically absorbed providing the required periodic feedback.
  Additional advantage: The period of the metallic grid can be adjusted according to the laser properties which first can be evaluated after the epitaxy.

Advantage of all these DFB concepts:

- The laser emission wavelength is given by the periodicity of the grating. Different longitudinal modes are more distant. Therefore, such a laser has a "single longitudinal mode" emission characteristics.
- In consequence, the emission wavelength gets more stable while mode hopping may occur in a Fabry–Pérot laser with a very dense longitudinal mode spectrum when the laser current is varied.
- Also the temperature stability is improved, as the wavelength change is now given by the weak temperature dependence of the refractive index instead of the stronger dependence of the band gap (see Eq. 1.29).
- In principle, the characteristic temperature (Eq. 12.18) can be maximized (or even driven into the negative region) by compensating the temperature dependent losses by an

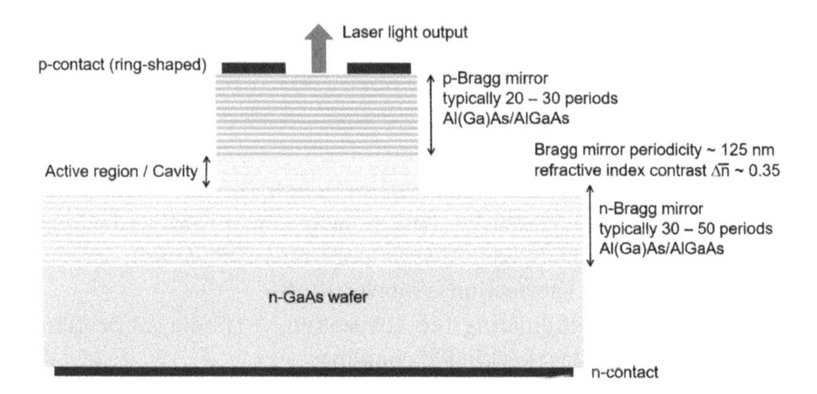

Figure 12.11 Vertical-cavity surface-emitting laser (schematically).

adequately de-tuned grating, where the tuning ("overlap" of optical gain curve and Bragg resonance of the grating) gets better with increasing temperature.

Such DFB lasers are key elements in optical long-distance data communication via glass fibers, where extremely narrow and stable laser lines are required for increasing the data transmission rates by wavelength division multiplexing methods.

12.2.9 Vertical-Cavity Surface-Emitting Laser

A fairly simple realization of a DBR laser is the so-called **VCSEL** (= vertical-cavity surface-emitting laser). Here, the Bragg gratings at either side of the cavity are grown epitaxially (typically by an Al(Ga)As-GaAs multi-heterostructure; c.f. Fig. 12.11). Hence, a fairly large refractive index step can be achieved, moreover, the grating period can be precisely controlled by the epitaxial process.

However, this concept requires that the light travels perpendicular to the interface planes, i.e., parallel to the growth direction. Hence, the interaction with the gain-providing active region is much shorter (several 10 nm) as compared to edge emitters (where the distance between the Fabry–Pérot mirrors, i.e., the cavity length is typically several 100 μm).

In order to get the same overall gain, the light has to pass more often the active region. This can be realized by a very high reflectivity

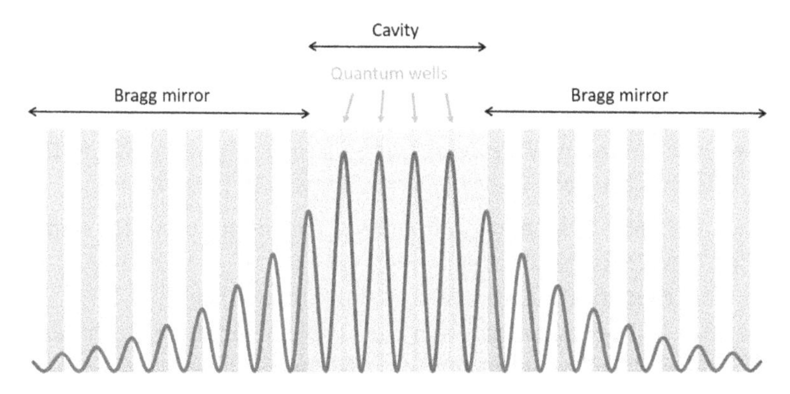

Figure 12.12 Field distribution of electric component of the laser light in a vertical-cavity surface-emitting laser (schematically).

of the Bragg mirrors. In order to get a similar optical path, the photons should be reflected about 1000 times before they leave the cavity. This requires mirrors with reflectivities far above 99%, i.e., very close to 100%.

Fortunately, this is not a big problem (at least for GaAs-based VCSELs) and can be achieved with 30–40 periods of an AlGaAs-Al(Ga)As heterostructure.

Anyway, the various layers of a VCSEL must precisely fulfill at least the following design considerations (see Fig. 12.12):

- Periodicity P of both Bragg grating must fit to the same wavelength λ: $P = (\lambda/4\bar{n}_1) + (\lambda/4\bar{n}_2)$ with $\bar{n}_i$ = refractive index of the two layer materials of the Bragg mirror (e.g., $Al_{0.95}Ga_{0.05}As$ and $Al_{0.5}Ga_{0.5}As$).
- The effective band gap of the semiconductor quantum wells must provide the maximum gain at exactly this wavelength.
- The distance of the Bragg mirrors, i.e., the width of the cavity must fit to a standing wave between the two Bragg mirrors.
- The positions of the quantum wells in the active region must be in the maxima of this standing wave.

All four requirements must be fulfilled simultaneously, which is a challenge for the epitaxial growth.

Today, this can be routinely done particularly for GaAs-based VCSELS with AlGaAs-AlGaAs Bragg mirrors. As mentioned before, this material system has the inherent advantage of a very low lattice mismatch. Such devices have the following advantages as compared to edge-emitting laser diodes:

- Once the epitaxial structure is grown (on a large area wafer), the devices can be processed in a similar way as LEDs, i.e.,
    - Cheap fabrication and contact formation, perfectly suited for mass production;
    - No critical cleaving of laser mirrors necessary;
    - Test of devices on wafer is possible before time-consuming and expensive single chip separation.

  Therefore, VCSELs are extremely cheap devices, similar as cheap LEDs.
- VCSELs can be fabricated with small diameters. Hence, extremely low threshold currents $I_{thr} < 100\ \mu A$ can be achieved. Notice: The active device area is in the range of a few $(\mu m)^2$, whereas for edge emitters with a length of several 100 μm and a width of at least 1–2 μm, it is significantly larger. However, the threshold current density is typically larger than in edge emitters.
- VCSELs typically have a circular shape, i.e., their output beam has a circular beam profile which is very advantageous for many applications, e.g., light coupling into a glass fiber.
- VCSELs can be arranged in arrays (one-dimensional or two-dimensional) quite easily. This may be applied in parallel fiber data communication.

However, VCSELs also have some inherent problems:

- For the Bragg mirrors, a large number of layer pairs are needed to fulfill the requirement of high reflectivity. This leads to a high electrical series resistance, in particular on the p-side of the device, requiring thorough optimization. Some improvement can be achieved by graded composition changes in the Bragg grating structures.
- The light output is strongest where the current density is strongest. However, as both have the same direction, the electrical contacts would impede the outcoupling of light. Typically, this problem is solved by forming a ring contact

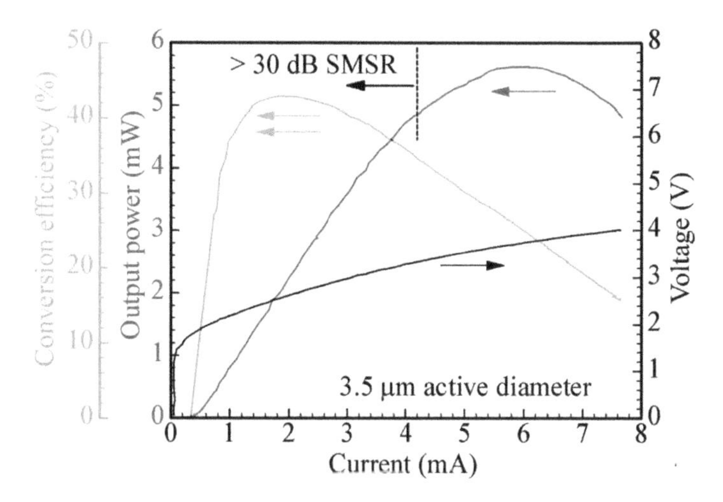

Figure 12.13 Typical voltage and output power of a GaAs vertical-cavity surface-emitting laser emitting at 840 nm (from [87], reproduced by permission of the Institution of Engineering & Technology).

(on one side) and a buried current aperture which drives the current into the center of the VCSEL mesa

- by making the outer part highly resistive by proton ion implantation
- by oxidation of an AlAs current aperture layer close to the active layers

- Due to the comparably small area (see above) the output power of a VCSEL is very low (some mW, Fig. 12.13). For larger diameters, the current confinement problems increase.

Over the last decades, GaAs-based VCSELs have found huge fields of application, including short-range optical data communication (e.g., replacing electrical connections in computer clusters), gas sensing, light source for high-resolution computer mice, etc. VCSELs made of other materials have been developed for other spectral ranges from infrared (e.g., 1.5 μm) over red to blue light-emitting devices. A very exhaustive description of VCSEL fundamentals, specific details, and applications can be found in Ref. [88].

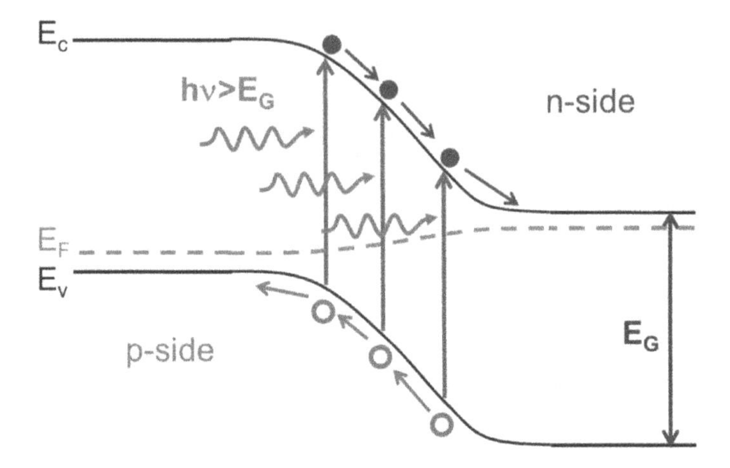

Figure 12.14 Working principle of a solar cell (schematically).

12.3 Solar Cells

Nowadays, semiconductor-based solar cells have gained strong attention as a source for renewable energy. Being fairly simple semiconductor devices, they can transform directly sun light into electrical current. In this chapter we will mainly focus on the application of compound semiconductors in such photovoltaic devices. Please notice: A photodiode which is used to detect light in specific applications (see, e.g., Section 6.4) has basically the same functionality.

12.3.1 Basic Working Principle

In principle, each semiconductor pn junction can act as a photovoltaic device (Fig. 12.14). When a photon is absorbed in the depletion region of the pn junction, an electron–hole pair is generated (c.f. Fig. 1.16 and Chapter 6). Without externally applied bias, a built-in voltage can be found in the depletion region as a consequence of the Fermi level position being close to the conduction band edge on the n-side and close to the valence band edge on the p-side of the junction. This is equivalent to an internal electric field from the n-side to the p-side. Hence the electrons will move to the n-side (they "fall down-hill"), while the holes move to the p-side[c]. Evidently, the built-in field helps to separate the generated carriers before they would recombine spontaneously. Under steady-state illumination, the carriers move to their respective contacts: A DC current is generated.

This behavior can be modeled by adding a current generation term j_{PV} to the basic diode equation

$$j_{\text{total}} = j_s \cdot (e^{\frac{qV}{nkT}} - 1) - j_{PV} \approx j_s \cdot e^{\frac{qV}{nkT}} - j_{PV}, \qquad (12.20)$$

where j_{total} is the total current density through the pn junction at a given voltage V and temperature T. As usual, k is the Boltzmann constant and q the elementary charge, while j_s is the diode saturation current. n represents the diode ideality factor, which is between 1 and 2. Of course, this equation can be formulated in the same way for the current I instead of the current density j.

This means: By illuminating a pn junction, the I–V curve is shifted downward by j_{PV}, see Fig. 12.15. The generated electrical power is the best product of the current and voltage in the lower right quadrant of Fig. 12.15. This so-called "MPP" (maximum power point) depends on the details of the I–V curve. It reaches best values for a maximized filling factor

$$FF = \frac{I_{\text{MPP}} \cdot V_{\text{MPP}}}{I_{\text{SC}} \cdot V_{\text{OC}}}, \qquad (12.21)$$

with the short-circuit current I_{SC} at $V = 0$ and the open-circuit voltage V_{OC} at $I = 0$.

It is clear, that such a photovoltaic current and power can only be generated by photons with an energy larger than the band gap in the semiconductor, i.e., in a solar cell, only this part of the solar spectrum can be "harvested," which fulfills this boundary condition. All photons with $h\nu < E_g$ are lost for this process.

On the other hand, if a photon with an energy $h\nu$ above the band gap E_g is absorbed, electrons and holes are excited to higher (resp.

---

[c]Please keep in mind that all conventionally used band structure pictures including those of this book are drawn for electrons: These particles have the lowest energy state when they are at the bottom of such diagrams, while the positively charged holes have their lowest energy state at the top-most position.

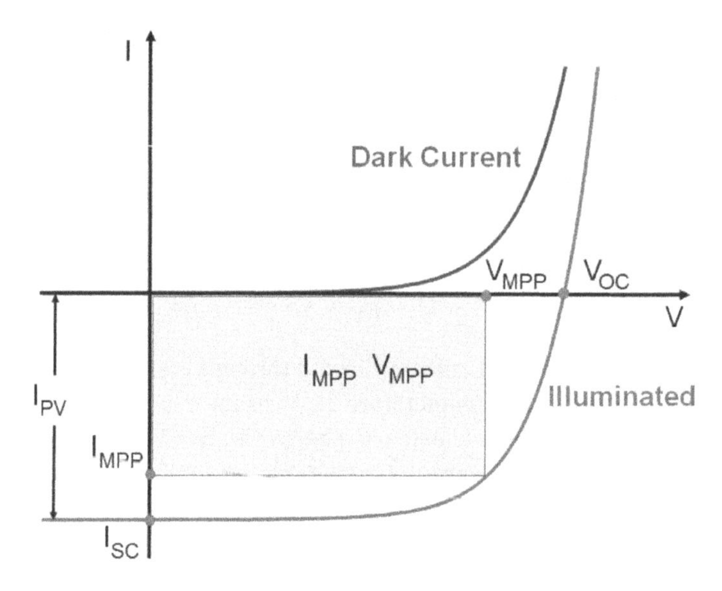

Figure 12.15 I–V curve of a single junction solar cell schematically.

lower) states in the conduction and valence band. As discussed in Chapter 6, they quickly relax to the respective band edge before getting extracted. Hence, these photons create finally an electron–hole pair separated (roughly) by the band gap energy, whereas the excess energy of the photon ($h\nu - E_g$) is typically transferred to phonons in the fast relaxation process, i.e., lost as heat.

Therefore, the maximum obtainable efficiency of such a single junction solar cell is far below 100%, depending on the details of the solar spectrum (see Section 12.3.2), as originally calculated by Shokley and Queisser in 1961 [89] (Fig. 12.16).

It is interesting to note that Si is still the mostly used material for solar cells owing to its abundant availability in excellent quality and low price, although it has a low absorption coefficient owing to its indirect band structure. In fact, the low absorption coefficient can be easily compensated by a thick absorber layer. Then, the indirect band structure helps to suppress the spontaneous recombination of the generated carriers on their way to the contacts; hence, they can move over long distances, as required by the thick absorber layers.

12.3.2 Solar Spectrum

The solar spectrum which eventually hits the photovoltaic cell differs according to many boundary conditions. The major factor is the earth's atmosphere and hence the path on which the light travels to the surface depending on the inclination of the sun. In order to have well-defined conditions for the characterization of solar cells, some spectral details are fixed by indicating an "air mass." Roughly speaking, the air mass coefficient (AM) is given by the sun's inclination:

$$AM = \frac{L}{L_0} = \frac{1}{\cos\varphi}, \tag{12.22}$$

where L is the path length of the sun light for a given inclination angle φ and L_0 is the zenith path length (normal to the earth's surface) at sea level.

Hence $AM1$ refers to conditions with vertical sunlight, while a very popular spectrum is the $AM1.5$ spectrum related to an inclination angle of about 48° (Fig. 12.16), as a large part of the earth's population lives in regions around this degree of latitude. For direct sunlight at noon, this results in an intensity slightly below 1 kW/m^2.

The solar spectrum near our planet in space, without any atmosphere, is indicated as $AM0$. Now, the intensity is about 1.353 kW/m^2.

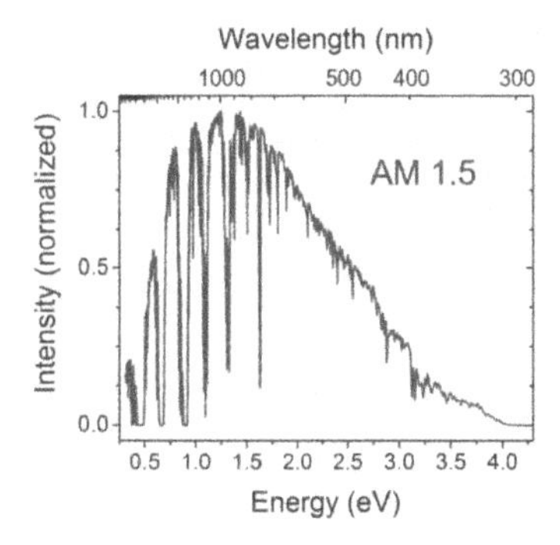

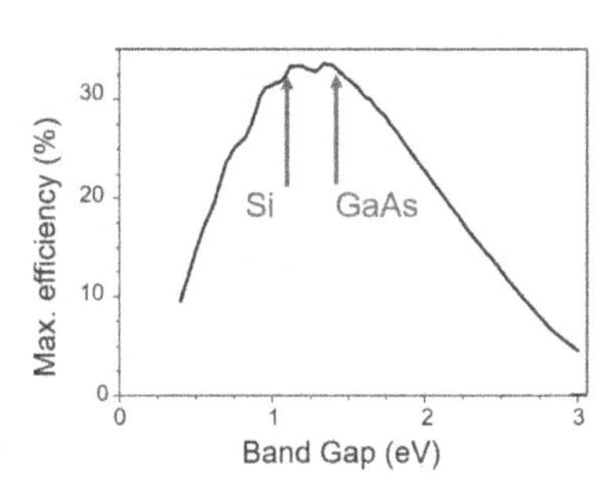

Figure 12.16 AM 1.5 spectrum of our sun (see Section 12.3.2) and maximum efficiency of semiconductor-based single-junction solar cells as function of band gap for AM1.5 illumination.

Please notice that solar cell efficiencies depend on the detailed shape of the spectra. Hence, some solar cells may have a higher efficiency under $AM1.5$ conditions as compared to $AM0$, which must be considered when discussing efficiency numbers.

12.3.3 Compound Semiconductors and Photovoltaics

Compound semiconductors have several obvious advantages as compared to Si for applications in solar cells:

- Due to their direct band structure (which is valid for most compound semiconductors, as discussed earlier in this book), only thin layers are required to absorb all suitable light (with $h\nu > E_g$). This becomes particularly important for solar cells for space applications, where the weight of the devices would critically contribute to the total weight to be transported into space.
- The large choice of compound semiconductors with various band gaps helps to find a material which provides best efficiency according to the Shockley–Queisser limit (Fig. 12.16).
- Compound semiconductors like GaAs are more stable against high energy particle radiation which is the main reason for solar cell degradation in space.
- The chance to form appropriate heterostructures can be used to cover the front side of a photovoltaic cell with a so-called window layer with higher band gap than the pn junction. This prevents minority carriers from diffusing to the surface of the device and recombining there. The same "trick" can be applied on the back-side of the device, creating a "back-surface field" repelling again the minority carriers from the back-side.
- Moreover, multi-junction devices can be realized, as discussed in more detail in the next section.

However, significant disadvantages still prevent their widespread use:

- The synthesis of compound semiconductors is by far more complicated and expensive than that of Si. This is particularly true for solar cells, which require huge areas for generating significant power from the sun light. For space applications, the price of the devices may play only a minor role compared to the required effort for transporting the devices into the orbit and for their long term stability with highest efficiency. However, for terrestrial applications, the device fabrication costs significantly change the game.

- Another issue which is heavily discussed, although still with somewhat unclear result is the potential toxicity of materials like GaAs, which would have a band gap close to the optimum for the solar spectrum. Although being a stable compound under normal conditions, it might be a contamination risk in case of accidents, e.g., fire.

12.3.4 Multi-Junction Solar Cell

Obviously, the combination of several cells made of material with different band gap would help to overcome the Shockley–Queisser limit. Such cells can be stacked, i.e., the top cell with large band gap E_{g1} would absorb the short wavelength part of the solar spectrum, letting pass all light with $h\nu < E_{g1}$. This light arrives on the next cell in the stack with band gap E_{g2} which absorbs photons of lower energy with $E_{g1} > h\nu > E_{g2}$. The still passing light with $h\nu < E_{g2}$ could then be absorbed in a third cell and so on.

In order to simplify their fabrication and later their operation, monolithically stacked multi-junction cells have been developed with up to 5 junctions. For this purpose, III-V compounds are well suited, as for each required band gap, a binary or ternary material can be found. Over the recent years, the epitaxial process for such multi-junction solar cells has been intensively studied. For each number of junctions, the optimum band gaps of the subcells can be determined. Several challenges in realizing such monolithic multi-junction stacks have been identified:

- A basic requirement for a defect-poor multi-heterostructure containing various materials is that all layers should have the same lattice constant. However, when checking the available materials (see Fig. 2.3), we notice that high-band gap materials have a smaller lattice constant than small band gap materials.

- When growing one pn junction on another junction, a connecting np junction is formed. This junction has a large electrical resistivity. It may even act as another photovoltaic cell switched reversely to the other two junctions hence decreasing the total PV current.
- All subcells are switched in series, i.e., the same current must flow through all junctions. On the other hand, it would be nice to optimize each subcell independently with regard to current and voltage, the optimum values being even sensitive to the spectrum and intensity (think about the solar spectrum at noon and at sun rise or sun set).

All these problems can be solved to some extent:

In modern multi-junction solar cells—three-junction cells are fairly standard, four- and five-junction cells are studied in the research labs—typically Ge is taken as substrate which, besides being cheap, has a lattice constant fairly well-suited to GaAs. Then, the low-band-gap subcells, having a larger lattice constant, are grown metamorphically on this substrate, before the higher band gap top cells are grown with the GaAs lattice constant. In some cases, the structures are grown in the reverse sequence, which eventually requires the removal of the substrate (GaAs or Ge) from the finally topmost high-band gap solar cell. Having carefully optimized the epitaxial processes—such structures are typically grown by MOVPE (c.f. Section 4.3)—these multi-junction structures have obtained an impressive maturity (see Section 12.3.6). Notice the need to grow more than about 30 different layers, all with perfect characteristics, for such monolithic stacks.

For semiconductor physicists, the solution to the next problem is obvious: The connecting np-junction must be a tunnel diode. In a classical tunnel diode, both, n-side and p-side are heavily doped (to levels above 10^{19} cm^{-3}) moving the Fermi level into the conduction and valence band, respectively. Moreover, the pn junction must be abrupt on a scale of few Angstrom. Then, for small applied voltage, states filled with electrons in the conduction band on one side are

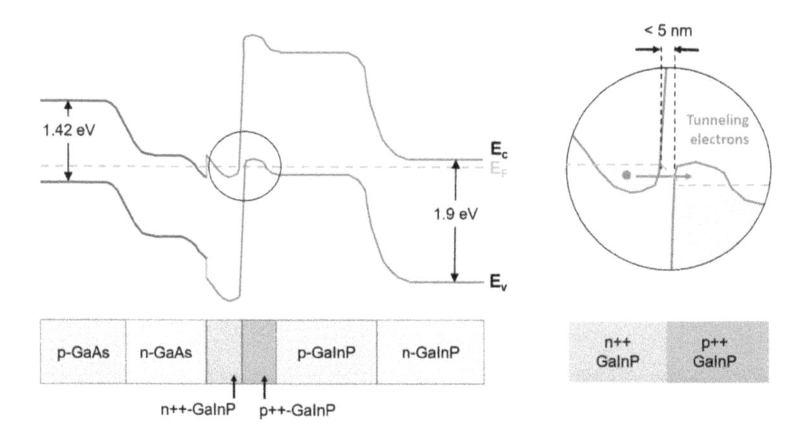

Figure 12.17 Schematic band diagram of a tandem solar cell made of a GaAs and a GaInP subcell, connected by a tunnel junction (in the region marked by the circle). The tunnel junction is enlarged in the right part of the diagram, now displayed with some bias. The shaded areas represent the states in the conduction band (left of junction) and valence band (right of junction) filled with electrons.

on the same energetic level as empty states (or states heavily filled with holes) on the other side. Due to the very small distance, the electrons can penetrate the (horizontal) gap by tunneling. Hence, a good tunnel junction has a very low resistance around $V = 0$ (see Fig. 12.17) for both directions of applied voltage.

The remaining challenge is the realization of such tunnel junctions for the appropriate materials in a multi-junction device, keeping in mind that particularly strong p-type doping with abrupt profile is not easy due to the diffusivity of nearly all acceptors and other, epitaxy-related problems like memory effects of the dopants, all leading to a less abrupt interface. Anyway, sufficient tunnel junction properties could be realized for the most interesting materials in terms of feasibility and optimum band gap.

Finally, current matching must be fulfilled: Each cell should generate the same current, still working in its optimum operating point. This can be achieved by an appropriate design of each subcell: The current generated by each subcell is (roughly) given by the amount of absorbed light, hence the thickness of the absorbing layers can be used as tuning parameter. However, this optimum depends on the solar spectrum, as discussed above. Therefore, in order to optimize a stack of junctions with respect to various illumination scenarios (e.g., morning, noon, evening, summer and winter), a best compromise needs to be found.

12.3.5 Concentrator Cells

Owing to the high price of compound semiconductor materials and device fabrication, it makes sense to minimize the area of the solar cells. The photovoltaic performance can be kept (or even improved, as discussed in the following) by concentrating the sunlight by lenses or other optical projection tools. In Eq. 12.20, the photovoltaic current density j_{PV} increases fairly linearly with the sunlight intensity.

Interestingly, the total efficiency, i.e., the generated electrical power with respect to the total optical intensity of the solar cell does not remain constant, as we may assume naively. From the photovoltaic diode equation 12.20, we can deduce for $I = 0$:

$$V_{\mathrm{OC}} = nkT \ln \frac{I_{\mathrm{PV}}}{I_{\mathrm{s}}} \tag{12.23}$$

i.e., the open-circuit voltage V_{OC} increases with increasing photo-generated current. Consequently, the total photo-generated power increases super-linearly, i.e., the total efficiency increases more or less logarithmically with I_{PV}, i.e., the sunlight concentration factor.

However, as usual, this advantage is not "free of charge." For sunlight concentration, some light projection devices like lenses or so are required. It is also clear that only the direct sunlight can be concentrated with good efficiency. Hence, concentrator systems only work properly if the optical system is accurately tracked with respect to the position of the sun during its course. They produce much less power from indirect sunlight from a cloudy sky as compared to standard systems.

Moreover, the absorbed light which is not converted into electrical power is transformed into heat. Consequently, concentrator systems typically require some cooling of the device. For very high concentration numbers, active cooling with water or other cooling fluids may be required to prevent an overheating and destruction of the cell.

Modern concentrator systems work with concentration levels between 200 and 1000, i.e., the light of 1 m^2 is concentrated onto a solar cell of only a few 10 cm^2.

12.3.6 PV Systems with Highest Efficiency

Obviously, highest efficiencies can be obtained by combining the technologies described in the last two subsections. Calculations demonstrate that efficiencies up to 53% can be achieved by triple junction solar cells, increasing to nearly 57% for four junctions [90], when each subcell is carefully designed, particularly regarding its individual band gap, for a sunlight concentration factor of 500. Currently, the best four-junction system has reached an efficiency of 46% at a sunlight concentration of 508 [91]. In order to overcome the problem of lattice mismatch between the lower band gap bottom junctions and the higher band gap top junctions, this structure was grown in two parts [92]: The two top junctions are grown lattice-matched on a GaAs substrate, starting with the higher band gap subcell made of $Ga_{0.51}In_{0.49}P$ (E_g = 1.88 eV), followed by a GaAs subcell (E_g = 1.42 eV). The two bottom junctions made of $Ga_{0.47}In_{0.53}As$ (E_g = 0.74 eV) and $Ga_{0.16}In_{0.84}As_{0.31}P_{0.69}$ (E_g = 1.12 eV) are grown lattice-matched on InP. They are connected by wafer-bonding, and finally the GaAs substrate is removed (Fig. 12.18).

In technologically simpler approaches, Ge is used as substrate—notice its similar lattice constant to GaAs, see Fig. 2.3—and also as low-band gap bottom cell, on which a metamorphic $Ga_{0.83}In_{0.17}As$ middle band gap cell is grown, covered by a $Ga_{0.35}In_{0.65}P$ large band gap cell. Such a three-junction solar cell has demonstrated an efficiency of more than 41% at a concentration of 454 suns [94].

Current research is concentrating on further optimizing the epitaxial growth of such multi-junction devices, preferably grown in one run on a single substrate. Hence problems related to the lattice mismatch between subcells with large and with small band

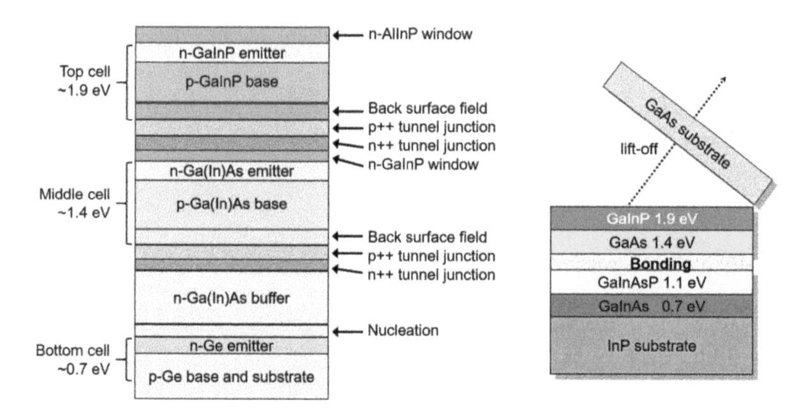

Figure 12.18 Layer sequence of highly efficient multi-junction solar cells. The left diagram (after [93]) shows a three-junction cell where the compound semiconductor heterostructure is grown monolithically fairly lattice-matched on a Ge substrate, the pn-junction in which forms the first subcell. On the right, the current record-holder four-junction solar cell is displayed [91], which was grown on two different substrates and then connected by wafer bonding. For simplicity, only the subcell materials are indicated.

gap are most prominent (find more details and discussions in [90]). One possibility is the growth of the full four-junction structure in inverted sequence on a GaAs substrate by starting with the lattice-matched top cells and growing a transition layer to the lattice-mismatched two bottom layers (metamorphic growth) [95]. Again, the substrate needs to be removed after epitaxy. Investigations for five-junction systems are ongoing.

Of course, the more junctions with different, but optimized band gap would be combined, the higher is the expected total efficiency. Ultimately, even for an infinite number of junctions, it will be limited by well-known thermodynamic laws, i.e., the Carnot heat engine efficiency taking the surface temperatures of the earth and of the sun as the lower and upper temperature. Analyzing the problem somewhat more in detail brings up more limiting arguments, see e.g., [96].

Problems

(1) (a) Explain how a simple light-emitting diode (LED) works.
(b) Why is it better to use a double heterostructure for an LED instead of a simple pn junction?
(c) What is (roughly) the minimum voltage needed to get light emission from an LED?
(d) What about the minimum current?

(2) Efficiency of a LED

(a) Explain the following terms:

- internal quantum efficiency
- injection efficiency
- extraction efficiency
- external quantum efficiency
- wall plug efficiency

(b) Discuss some typical phenomena which may influence these efficiencies.
(c) What is the difference between external quantum efficiency and wall plug efficiency? In which units are they measured?

(3) What is described by the terms "current crowding" and "current spreading"? Explain their relevance with respect to a LED.

(4) What is meant by the term "green gap"?

(5) Two LEDs are characterized by the same luminous efficiency of $\eta = 100\,\mathrm{lm/W}$. One LED emits green light, the other red light. Which one (if any) has a larger wall plug efficiency?

(6) How can white light be generated by LEDs? Discuss pros and cons of different approaches.

(7) (a) Explain how a laser diode works.
(b) Describe the first laser condition. How can it be fulfilled?
(c) What requirements lead to the second laser condition?
(d) What is the difference between the transparency current and the threshold current in a laser diode?
(e) What should you do with the laser facets to decrease the threshold current of the device?

(8) Discuss the major technological advances which eventually led to the actual state of excellent laser diodes.
(9) Explain the need and the function of "separate confinement."
(10) Discuss pros and cons of varying the confinement factor in a modern laser diode.
(11) (a) What is described by the term "characteristic temperature" for a laser diode?
(b) How can you improve the characteristic temperature in laser diodes emitting at the same wavelength?
(12) How can a lateral wave guiding be achieved in a laser diode?
(13) Explain the terms "near field" and "far field."
(14) Bragg gratings and distributed feedback:

(a) Describe the difference between a conventional Fabry–Pérot laser and a DFB laser.
(b) What is the advantage of a DFB laser?
(c) What is the main problem in fabricating a DFB laser?
(d) How does the emission wavelength of a DFB laser change with temperature as compared to a Fabry–Pérot laser? Why?
(e) How can you design a laser diode with negative characteristic temperature for the threshold current (in some temperature range, e.g., around room temperature)?

(15) Vertical-cavity surface-emitting lasers:

(a) Explain the advantages of a VCSEL as compared to edge-emitting laser diodes.
(b) What are the technological challenges of preparing a VCSEL?
(c) What are the inherent problems of a VCSEL? Discuss solutions how to minimize such problems.

(16) Explain how a simple solar cell works. Why is a pn junction needed?
(17) Try to estimate (by simple assumptions) which is the optimum band gap of a semiconductor for a solar cell working with black-body radiation light of $T = 5000\,\mathrm{K}$.
(18) Find I_{MPP} and V_{MPP} for a solar cell with

$$I_{\mathrm{total}} = I_{\mathrm{s}} \cdot e^{\frac{qV}{nkT}} - I_{\mathrm{PV}}$$

(19) Estimate the required thickness for a solar cell made of GaAs. Compare to Si.
(20) Explain the functionality of a tunnel diode. Why is it well suited as connection between two stacked solar cells?
(21) Why is Ge an attractive substrate for multi-junction solar cells?
(22) Concentrator solar cells:
   (a) How do the current, voltage, and efficiency of a solar cell change with increasing concentration of the sunlight?
   (b) What will limit the efficiency for very high concentrations?
   (c) What other technical features/needs may make a concentrator solar cell system less attractive?

Chapter 13

Electronic Devices

Electronic devices like diodes, transistors, and integrated electronic circuits are mostly made of silicon. Here, we assume that the reader is familiar with the basic knowledge of the functionality of these devices, which can be found in many textbooks, e.g., refs. [97, 98]. However, some properties of the III-V compound semiconductors are very advantageous for such devices, in particular,

- wide choice of band gaps, in particular large band gap for high voltage high power devices;
- small effective electron masses in many compounds, which is excellent for high-frequency applications;
- formation of complex heterojunctions possible, which enables new device concepts.

Here, we will discuss a few concepts of electronic devices with special focus on the features specific for compound semiconductors.

13.1 Field-Effect Transistor

In Si electronics, the field-effect transistor (FET) is one of the most successful transistor concepts (see Fig. 13.1).

Compound Semiconductors: Physics, Technology, and Device Concepts
Ferdinand Scholz

ISBN 978-981-4774-07-9 (Hardcover), 978-1-315-22931-7 (eBook)
www.panstanford.com

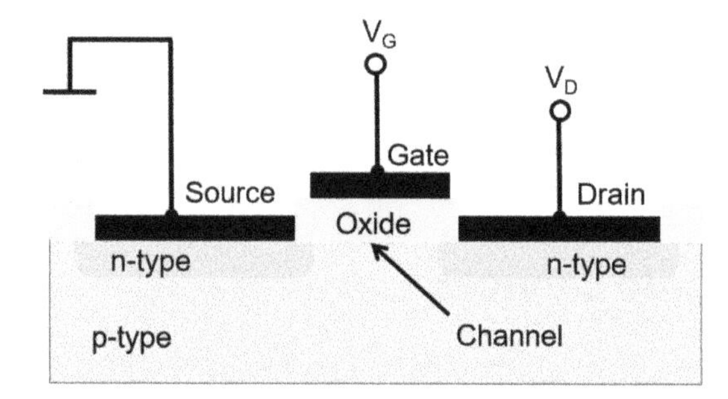

Figure 13.1 Metal-oxide-semiconductor field-effect transistor (schematically).

A current between the two electrodes **drain** and **source** is controlled by a third electrode, the **gate**, which allows to change the band alignment near the surface and hence the conductivity of the respective carrier channel. The gate is isolated from the semiconductor by a native SiO_2 layer, hence the name **M**etal **O**xide **S**emiconductor FET (MOSFET).

Different from Si, compound semiconductors do not form a suitable native oxide which could be used as a gate isolation layer. Different concepts have been developed:

- MESFET: MEtal Semiconductor FET: The gate electrode is realized with a metal forming a Schottky contact, so that no isolation is needed. Otherwise, this transistor is very much the same as a Si MOSFET.
- MISFET: Metal Insulator Semiconductor FET: Here, an artificial insulation layer is placed below the gate contact, typically SiO_2 or Si_xN_y.

More abbreviations are used in order to distinguish some different device concepts:

- MODFET = Modulation-doped FET: Just a thin doped layer is grown close to the surface.
- HFET = Heterostructure FET: On top of the lower-band gap material (e.g., GaAs) a thin layer (about 30 nm) of another material with higher band gap (e.g., AlGaAs) is grown. Now, the

gate controls the carrier concentration close to the interface instead near the surface.

- TEGFET = Two-dimensional Electron Gas FET, or HEMT = High Electron Mobility Transistor: As discussed in earlier chapters (see, e.g., Fig. 2.4), a two-dimensional electron gas may form at interfaces like the AlGaAs-GaAs interface, in which the electrons have a higher in-plane mobility because
  - Carriers are separated from dopant impurities (the latter being in the barrier material)
  - Less scattering processes due to screening effects as a consequence of the high carrier concentration,
  - Less scattering due to the reduced dimensionality.

  Notice:
  - In a GaAs MESFET the electrons have typically mobilities of about 2000–3000 cm^2/Vs, whereas
  - in a HEMT device, the mobility reaches values of 8000–9000 cm^2/Vs.
- The mobility can be further enhanced by biaxial strain in a GaInAs-(Al)GaAs p-HEMT (pseudomorphically strained HEMT) owing to the reduced electron effective mass (see Chapter 9).

Eventually, not the mobility determines the ultimate speed, but the so-called "saturation velocity": While for small fields, the carrier velocity v_G typically increases linearly with the applied electric field E, governed by the carrier mobility μ according to Eq. 5.6, it saturates for larger fields (Fig. 13.2). Here, some compound semiconductors provide better data than Si or Ge (Table 13.1). Moreover, some possess a higher internal break-down field, which means, that a larger voltage can be applied to the device before it breaks through. Consequently, such materials are well suited for high voltage, i.e., high power high frequency applications, e.g., in the output stage of mobile phones.

Another special situation is realized in the material system GaN-AlGaN. Here, opposite electric fields develop in GaN and AlGaN due to the spontaneous and piezoelectric properties, as discussed in Section 11.6. Similar as in a GaAs-AlGaAs-based HFET

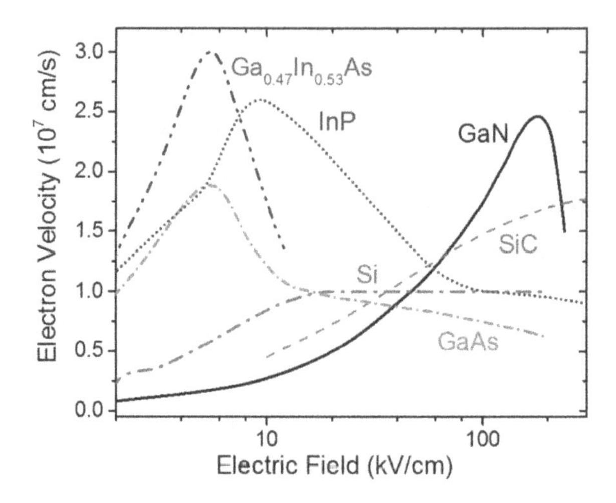

Figure 13.2 Theoretically predicted electron drift velocity versus applied electrical field for some compound semiconductors. Data from [98–101].

Table 13.1 Peak saturation velocity of some compound semiconductors and Si

Material	Peak saturation velocity (10^7 cm/s)	Electric field at peak (kV/cm)	Ref.
InSb	5.5	0.5 (at 77 K)	[102]
InAs	2.8	3	[103]
GaAs	1.8	3	[104]
InP	2.6	10	[99]
GaN	2.5	180	[100]
Si	1	>100	[105]

or HEMT as discussed above, a triangular quantum well forms at the interface. However, here, the resulting polarization field gives rise to the generation of two-dimensional free electrons with a typical concentration of $1 - 2 \times 10^{13}$ cm^{-2} at the interface with enhanced mobility (1000–1800 cm^2/Vs) even without any doping (Fig. 13.3), whereas in GaAs-based HEMTs, doping induced carrier concentrations range below about 5×10^{12} cm^{-2}.

Other advantages of this material system are

- extremely high electron saturation velocity (2.5×10^7 cm/s);

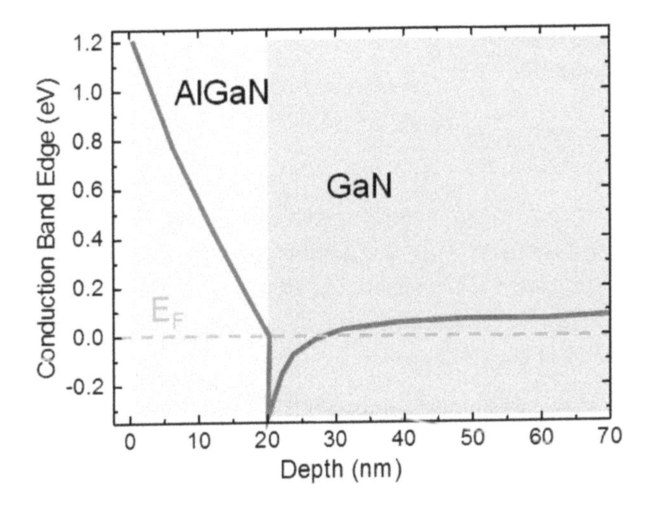

Figure 13.3 Band structure of a GaN-AlGaN FET structure after data from [106]. The strong electric field in the AlGaN barrier is due to the spontaneous and piezoelectric polarization (see Section 11.6). Even without intentional doping, a two-dimensional carrier gas at the interface is formed.

- high breakdown voltage, as the material can withstand electric fields of at least 3 MV/cm;
- high operating temperature due to the high band gap and the high thermal stability of GaN.

A major issue for such GaN-based FETs is still the substrate. Here, SiC is preferred because of its excellent thermal conductivity, which helps to get rid of the heat generated by Joule's losses in such high-power devices. Also, GaN-based HEMTs grown on Si perform quite well, whereas sapphire is much less suited due to its low thermal conductivity.

13.2 Heterobipolar Transistor

In a bipolar transistor, the current from the **emitter** to the **collector** I_C is controlled by the voltage applied to the **base** electrode. Typically, the respective currents are discussed.

The current gain β of this type of transistor (in npn configuration) is described by [107]

$$\beta = \frac{\partial I_C}{\partial I_B} = \frac{D_n}{D_p} \cdot \frac{w_E}{w_B} \cdot \frac{n_B}{p_E} = \frac{D_n}{D_p} \cdot \frac{w_E}{w_B} \cdot \frac{N_E}{N_B} \cdot \frac{e^{\frac{E_{g,E}}{kT}}}{e^{\frac{E_{g,B}}{kT}}}, \qquad (13.1)$$

where D_i is the diffusion length of the carriers, w_i is the width of the respective zone, and N_i is the doping concentration in the emitter (E) and the base (B), respectively. Here we made use of the mass action law (Eq. 1.42) to determine the minority carriers in the base (n_B) and in the emitter (p_E):

$$n_B = \frac{n_{i,B}^2}{N_B} \quad \text{with} \quad n_i \sim e^{-\frac{E_g}{2kT}}$$

In a homojunction, where the band gaps $E_{g,i}$ are the same in both zones, it is good to make N_E/N_B large. This obviously reduces the hole current out of the base which is a parasitic current in a npn transistor. However, for a low Ohmic contact to the base, the base should be heavily doped.

In a heterojunction bipolar transistor (HBT, see Fig. 13.4), another degree of freedom exists:

Make $\Delta E_g = E_{g,E} - E_{g,B}$ large, because

$$\beta \sim \frac{N_E}{N_B} \cdot e^{\frac{E_{g,E}-E_{g,B}}{kT}}, \qquad (13.2)$$

One can easily gain a factor of 10,000 and more for $\Delta E_g \sim 10kT$! Now, high p-type doping of the base is possible.

Additionally, the specific band line-up should be optimized:

- For lower ΔE_C, the electron current suffers less from this heterobarrier, whereas
- A larger ΔE_V permits higher p-type doping in the base.

Therefore, the system GaInAs-InP with $\Delta E_C/\Delta E_V = 42/58$ is better than GaInAs-AlInAs (67/33), and GaInP/GaAs (25/75) is better than AlGaAs-GaAs (65/35).

The remaining step in the conduction band may be further decreased by a composition grading.

We see phosphide-arsenide heterostructures are better than pure arsenides. However, it is more difficult to grow epitaxially high

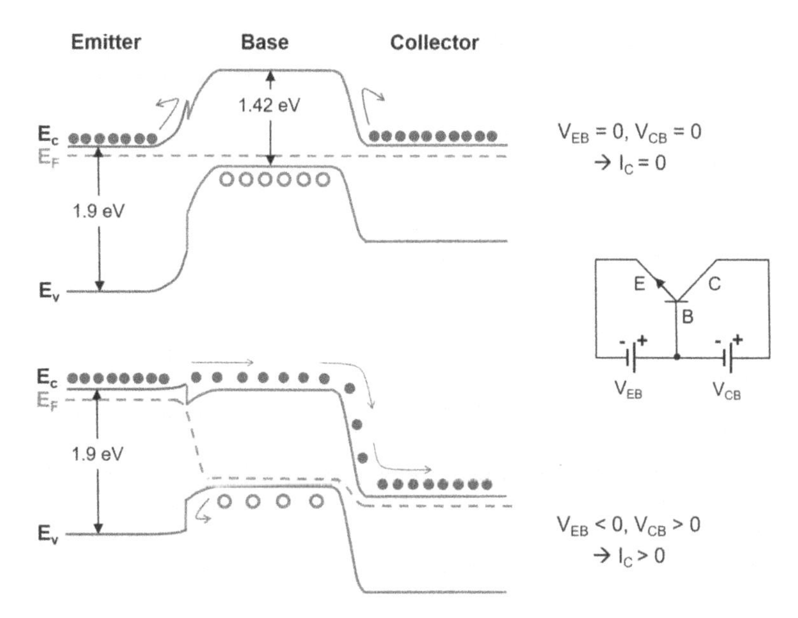

Figure 13.4 Naive band diagram of a heterobipolar transistor structure without (top) and with applied bias (bottom). Note the larger band gap in the emitter region. The barrier in the valence band blocks the holes to diffuse from the base to the emitter region.

quality P-As interfaces. This is in particular difficult in MBE, which has problems in handling P.

In order to achieve a highly p-doped base (which is required for low base resistance and hence high frequency devices), C doping instead of Be (or Zn) is applied to GaAs.

InP is particularly well suited as HBT substrate material, because of

- higher mobility of GaInAs enabling to achieve higher frequencies;
- low surface recombination velocity which is particularly important for realizing sub-micrometer devices;
- high thermal conductivity of InP (0.68 W/cm K; GaAs: 0.55 W/cm K);
- low 1/f-noise;
- radiation hardness (military, space applications);

- lower turn-on voltage, which leads to low power dissipation, a very important advantage for portable applications.

Indeed, amplifiers working up to and beyond several 100 GHz (e.g., $f_T = 600$ GHz; $f_{max} = 246$ GHz) have been reported [108].[a]

However, today, GaInP-GaAs HBTs are mainly produced as they use the cheaper GaAs as substrate.

Problems

(1) Explain with simple arguments how a triangular quantum well is formed at a GaAs–AlGaAs interface.

(2) What material properties make compound semiconductors well-suited for FETs?

(3) Field-effect transistor (FET):

   (a) Explain how the conductivity of the channel of a field effect transistor can be varied by changing the gate voltage.

   (b) What is the advantage of using an AlGaAs-GaAs FET instead of a Si FET?

   (c) What is the qualitative difference between an AlGaAs-GaAs FET and an AlGaN-GaN FET?

(4) Hetero-bipolar transistor (HBT):

   (a) Which part of a HBT should have a large band gap?

   (b) How should the band offsets be distributed best between conduction band and valence band?

   (c) Which part of a HBT is most critical with respect to doping? Explain.

[a] f_T = Current gain cutoff frequency (amplification > 1); f_{max} = maximum frequency of oscillation.

References

1. C. Kittel, *Introduction to Solid State Physics* (John Wiley, New York, 1996).
2. J. Wilson and J. Hawkes, *Optoelectronics* (Prentice Hall, Harlow, 1998).
3. Y. P. Varshni, Temperature dependence of the energy gap in semiconductors, *Physica* **34**, 149 (1967).
4. E. Rosencher and B. Vinter, *Optoelectronics* (Cambridge University Press, Cambridge, UK, 2002).
5. A. K. Saxena, The conduction band structure and deep levels in $Ga_{1-x}Al_xAs$ alloys from a high-pressure experiment, *J. Phys. C: Solid State Phys.* **13**, 4323 (1980).
6. J. Chikawa and J. Matsui, in *Handbook on Semiconductors*, T. S. Moss, and M. Balkanski (ed.) (Elsevier Science, Amsterdam, 1994), vol. 3a, ch. 1, p. 1.
7. J. B. Mullin, in *III-V-Semiconductor Materials and Devices*, vol. 7 from *Materials Processing: Theory and Practices*, R. J. Malik (volume ed.), F. F. Y. Wang (series ed.) (Elsevier, Amsterdam, 1989), ch. 1, p. 1.
8. P. Rudolph and M. Jurisch, Bulk growth of GaAs: An overview, *J. Cryst. Growth* **198/199**, 325 (1999).
9. http://www.wafertech.co.uk/growth.html (2016) with permission of Wafer Technology Ltd, a U.K. based producer of III-V materials and epitaxy-ready substrates.
10. N. Tabatabaie, V. M. Robins, and G. E. Stillman, in *III-V-Semiconductor Materials and Devices*, vol. 7 from *Materials Processing: Theory and Practices*, R. J. Malik (volume ed.), F. F. Y. Wang (series ed.) (Elsevier, Amsterdam, 1989), ch. 2, p. 73.
11. M. G. Astles, *Liquid-Phase Epitaxial Growth of III-V Compound Semiconductor Materials and Their Device Applications* (Adam Hilger, Bristol, 1990).

12. H. Eisele, *Thermodynamische und physikalische Eigenschaften des ternären Systems InGaAs*, Ph.D. thesis, Universität Stuttgart, 1985.
13. W. Körber, *Epitaktischer Einbau und optische Eigenschaften seltener Erden in III-V-Halbleitern*, Ph.D. thesis, Universität Stuttgart, 1988.
14. K. Onabe, Unstable regions in III-V quaternary solid solutions composition plane calculated with strictly regular solution approximation, *Jpn. J. Appl. Phys.* **21**, L323 (1982).
15. G. B. Stringfellow, *Organometallic Vapor Phase Epitaxy*, 2nd Ed. (Academic Press, San Diego, 1999).
16. P. D. Dapkus and J. J. Coleman, in *III-V-Semiconductor Materials and Devices*, vol. 7 from *Materials Processing: Theory and Practices*, R. J. Malik (volume ed.), F. F. Y. Wang (series ed.) (Elsevier, Amsterdam, 1989), ch. 4, p. 147.
17. M. Razhegi, in *Handbook on Semiconductors*, T. S. Moss and M. Balkanski (ed.) (Elsevier Science, Amsterdam, 1994), vol. 3a, ch. 3, p. 183.
18. H. M. Manasevit and W. I. Simpson, The use of metal-organics in the preparation of semiconductor materials, *J. Electrochem. Soc.* **116**, 1725 (1969).
19. G. Stringfellow, Thermodynamic aspects of organometallic vapor phase epitaxy, *J. Cryst. Growth* **62**, 225 (1983).
20. G. Stringfellow, The role of impurities in III/V semiconductors grown by organometallic vapor phase epitaxy, *J. Cryst. Growth* **75**, 91 (1986).
21. D. Schmitz, G. Strauch, J. Knauf, H. Jürgensen, M. Heyen, and K. Wolter, Large area growth of extremely uniform AlGaAs/GaAs quantum well structures for laser applications by effective LP-MOVPE, *J. Cryst. Growth* **93**, 312 (1988).
22. P. M. Frijlink, J. L. Nicolas, and P. Suchet, Layer uniformity in a multiwafer MOVPE reactor for III-V compounds, *J. Cryst. Growth* **107**, 166 (1991).
23. B. Joyce, P. Dobson, J. Neave, K. Woodbridge, J. Zhang, P. Larsen, and B. Bôlger, RHEED studies of heterojunction and quantum well formation during MBE growth: from multiple scattering to band offsets, *Surf. Sci.* **168**, 423 (1986).
24. M. M. May, H.-J. Lewerenz, and T. Hannappel, Optical in situ study of InP(100) surface chemistry: dissociative adsorption of water and oxygen, *J. Phys. Chem. C* **118**, 19032 (2014).

25. S. Sasaki, in *III-V-Semiconductor Materials and Devices*, vol. 7 from *Materials Processing: Theory and Practices*, R. J. Malik (volume ed.), F. F. Y. Wang (series ed.) (ed.) (Elsevier, Amsterdam, 1989), ch. 5, p. 217.
26. E. H. Parker, *Molecular Beam Epitaxy* (Plenum Press, New York, 1981).
27. M. A. Herman and H. Sitter, *Molecular Beam Epitaxy: Fundamentals and Current Status*, vol. 7 from Springer Series in Materials Science (Springer, Berlin, 1996).
28. M. Henini (ed.), *Molecular Beam Epitaxy: From Research to Mass Production*, (Elsevier Science, Oxford, 2013).
29. A. Y. Cho and K. Y. Cheng, Growth of extremely uniform layers by rotating substrate holder with molecular beam epitaxy for applications to electro-optic and microwave devices, *Appl. Phys. Lett.* **38**, 360 (1981).
30. http://www.lesker.com/newweb/Vacuum_Pumps/vacuumpumps_technicalnotes_1.cfm (2010).
31. P. M. Thibado, G. J. Salamo, and Y. Baharav, Robust optical delivery system for measuring substrate temperature during molecular beam epitaxy, *J. Vac. Sci. Technol. B* **17**, 253 (1998).
32. M. B. Panish and H. Temkin, *Gas Source Molecular Beam Epitaxy*, vol. 26 from Springer Series in Materials Science (Springer, Berlin, 1993).
33. J. S. Foord, G. J. Davies, and W. T. Tsang (ed.), *Chemical Beam Epitaxy and Related Techniques* (John Wiley, Chichester, 1997).
34. C. M. Wolfe, G. E. Stillman, and W. T. Lindley, Electron mobility in high-purity GaAs, *J. Appl. Phys.* **41**, 3088 (1970).
35. L. J. van der Pauw, A method of measuring specific resistivity and Hall effect of discs of arbitrary shape, *Philips Research Reports* **13**, 1 (1958).
36. T. Gairing, *Charakterisierung epitaktisch hergestellter Halbleiterlaser-strukturen*, Diploma thesis, Universität Stuttgart, 1994.
37. R. G. Wilson, F. A. Stevie, and C. W. Magee, *Secondary Ion Mass Spectrometry* (Wiley, New York, 1989).
38. D. Briggs and M. P. Seah (ed.) *Practical Surface Analysis* (Salle & Sauerländer, Aarau, 1996), vol. 2: Ion and Neutral Spectroscopy.
39. M. Keßler, *Grundlagen und Anwendungen der Zinkdiffusion in GaInP-Hochleistungslasern*, Ph.D. thesis, Universität Stuttgart, Der andere Verlag, Osnabrück, 1988.
40. http://www.chemicool.com/definition/quadrupole_mass_spectrometry.html, from Chemicool Web page (2016). Accessed 8/25/2016.

41. P. Bhattacharya, *Semiconductor Optoelectronic Devices* (Prentice Hall, London, 1997).
42. P. Y. Yu and M. Cardona, *Fundamentals of Semiconductors* (Springer, Berlin, 1996).
43. K. Seeger, *Semiconductor Physics*, 9th Ed. (Springer, Berlin, 2004).
44. V. V. Kopyev, I. A. Prudaev, and I. S. Romanov, Comparative analysis of efficiency droop in InGaN/GaN light-emitting diodes for electrical and optical pumping conditions, *J. Phys.: Conf. Ser.* **541**, 012055 (2014).
45. R. J. Dieter, *Defekte bei der Heteroepitaxie von Galliumarsenid auf Silizium*, Ph.D. thesis, Universität Stuttgart (Shaker Verlag Aachen), 1993.
46. A. Moritz, *Optische Verstärkung in GaInP/AlGaInP Quantenstrukturen*, Ph.D. thesis, Universität Stuttgart (Shaker Verlag Aachen), 1997.
47. http://franklin.chm.colostate.edu/skohli/HRXRD.pdf (2011).
48. A. Krost, in *Nano-Optoelectronics*, M. Grundmann (ed.) (Springer, Berlin, 2002), ch. 6, p. 135.
49. *Strained Layer Superlattices*, vol. 32 and 33 from *Semiconductors and Semimetals*, R. K. Willardson and E. R. Weber (ed.) (Academic Press, San Diego, 1990).
50. W. W. Chow, S. W. Koch, and M. Sargent III, *Semiconductor-Laser Physics* (Springer, Berlin, 1994).
51. M. P. C. M. Krijn, Heterojunction band offsets and effective masses in III-V quaternary alloys, *Semicond. Sci. Technol.* **6**, 27 (1991).
52. C. G. van de Walle, Band lineups and deformation potentials in the model-solid theory, *Phys. Rev. B* **39**, 1871 (1989).
53. J. Chen, J. R. Sites, I. L. Spain, M. J. Hafich, and G. Y. Robinson, Band offset of GaAs/InGaP measured under hydrostatic pressure, *Appl. Phys. Lett.* **58**, 744 (1991).
54. G. E. Pikus and G. L. Bir, Effect of deformation on the hole energy spectrum of germanium and silicon, *Sov. Phys. - Solid State* **1**, 1502 (1959).
55. E. P. O'Reilly, Valence band engineering in strained-layer structures, *Semicond. Sci. Technol.* **4**, 121 (1989).
56. P. Lawaetz, Valence-Band Parameters in Cubic Semiconductors, *Phys. Rev. B* **4**, 3460 (1971).
57. J. M. Luttinger, Quantum theory of cyclotron resonance in semiconductors: general theory, *Phys. Rev* **102**, 1030 (1956).

58. J. W. Matthews and A. E. Blakeslee, Defects in epitaxial multilayers, *J. Cryst. Growth* **27**, 118 (1974).
59. D. Bimberg, M. Grundmann, and N. N. Ledentsov, *Quantum Dot Heterostructures* (Wiley, Chichester, 1998).
60. M. Grundmann (ed.), *Nano-Optoelectronics: Concepts, Physics and Devices, NanoScience and Technology* (Springer, Berlin, 2002).
61. D. Bimberg (ed.), *Semiconductor Nanostructures, NanoScience and Technology* (Springer, Berlin Heidelberg, 2008).
62. M. Asada, Y. Miyamoto, and Y. Suematsu, Gain and the threshold of three-dimensional quantum-box lasers, *IEEE J. Quantum Electron.* **22**, 1915 (1986).
63. Y. Arakawa and H. Sakaki, Multidimensional quantum well laser and temperature dependence of its threshold current, *Appl. Phys. Lett.* **40**, 939 (1982).
64. C. Constantin, E. Martinet, F. Lelarge, K. Leifer, A. Rudra, and E. Kapon, Influence of strain and quantum confinement on the optical properties of InGaAs/GaAs V-groove quantum wires, *J. Appl. Phys.* **88**, 141 (2000).
65. F. C. Frank and J. H. van der Merwe, One-dimensional dislocations. I. static theory, *Proc. Roy. Soc. London A* **198**, 205 (1949).
66. M. Volmer and A. Weber, Keimbildung in übersättigten Gebilden (Nucleation of supersaturated structures), *Z. Phys. Chem.* **119**, 277 (1926).
67. I. N. Stranski and L. Krastanow, Zur Theorie der orientierten Ausscheidung von Ionenkristallen aufeinander, *Sitzungsber. Kais. Akad. Wiss. Wien Math.-Naturwiss. Kl. 2b* **146**, 797 (1938).
68. S. V. Ghaisas and S. Das Sarma, Surface diffusion length under kinetic growth conditions, *Phys. Rev. B* **46**, 7308 (1992).
69. M. Geiger, *Selbstorganisationseffekte bei Quantenpunkten: Metallorganische Gasphasenepitaxie und optische Charakterisierung*, Ph.D. thesis, Universität Stuttgart (Shaker Verlag Aachen), 1998.
70. *GaN I and II*, vols. 50 and 57 from *Semiconductors and Semimetals*, R. K. Willardson and E. R. Weber (ed.) (Academic Press, San Diego, 1998/1999).
71. T. Yao and S.-K. Hong (ed.), *Oxide and Nitride Semiconductors*, Advances in Materials Research, (Springer, Berlin Heidelberg, 2009).
72. S. Pearton (ed.), *GaN and ZnO-based Materials and Devices*, Springer Series in Materials Science (Springer, Berlin Heidelberg, 2012).

73. R. Dwilinski, R. Doradzinski, J. Garczynski, L. P. Sierzputowski, A. Puchalski, Y. Kanbara, K. Yagi, H. Minakuchi, and H. Hayashi, Excellent crystallinity of truly bulk ammonothermal GaN, *J. Cryst. Growth* **310**, 3911 (2008).
74. D. Ehrentraut, Y. Kagamitani, T. Fukuda, F. Orito, S. Kawabata, K. Katano, and S. Terada, Reviewing recent developments in the acid ammonothermal crystal growth of gallium nitride, *J. Cryst. Growth* **310**, 3902 (2008).
75. X. Ke, W. Jian-Feng, and R. Guo-Qiang, Progress in bulk GaN growth, *Chinese Phys. B* **24**, 066105 (2015).
76. H. Amano, M. Kito, K. Hiramatsu, and I. Akasaki, P-type conduction in Mg-doped GaN treated with low-energy electron beam irradiation (LEEBI), *Jpn. J. Appl. Phys.* **28**, L2112 (1989).
77. T. Matsuoka, N. Yoshimoto, T. Sasaki, and A. Katsui, Wide-gap semiconductor InGaN and InGaAIN grown by MOVPE, *J. Electron. Mat.* **21**, 157 (1992).
78. S. Sze, *Physics of Semiconductor Devices* (John Wiley & Sons, New York, 1981).
79. S. L. Chuang, *Physics of Optoelectronic Devices* (John Wiley & Sons, New York, 1995).
80. http://www.ecse.rpi.edu/~schubert/Light-Emitting-Diodes-dot-org (2007).
81. J. J. Vos, Colorimetric and photometric properties of a 2° fundamental observer, *Color Res. Appl.* **3**, 125 (1978).
82. T. W. Murphy, Maximum spectral luminous efficacy of white light, *J. Appl. Phys.* **111**, 104909 (2012).
83. Y. Narukawa, M. Ichikawa, D. Sanga, M. Sano, and T. Mukai, White light emitting diodes with super-high luminous efficacy, *J. Phys. D: Appl. Phys.* **43**, 354002 (2010).
84. G. Winstel, and C. Weyrich (ed.), *Optoelektronik I* (Springer-Verlag, Berlin, 1981), vol. 50.
85. R. Kersten, *Einführung in die optische Nachrichtentechnik* (Springer, Berlin, 1983).
86. Z. Alferov, in *Nano-Optoelectronics*, M. Grundmann (ed.) (Springer, Berlin, 2002), ch. 1, p. 3.
87. C. Jung, R. Jäger, M. Grabherr, P. Schnitzer, R. Michalzik, B. Weigl, S. Müller and K. J. Ebeling, 4.8 mW single-mode oxide confined top-surface emitting vertical-cavity laser diodes, *Electron. Lett.* **33**, 1790 (1997).

88. R. Michalzik (ed.), *VCSELs: Fundamentals, Technology and Applications of Vertical-Cavity Surface-Emitting Lasers*, Springer Series in Optical Sciences (Springer, Berlin Heidelberg, 2012).
89. W. Shockley and H. J. Queisser, Detailed balance limit of efficiency of p-n junction solar cells, *J. Appl. Phys.* **32**, 510 (1961).
90. D. Friedman, Progress and challenges for next-generation high-efficiency multijunction solar cells, *Curr. Opin. Solid State Mater. Sci.* **14**, 131 (2010).
91. F. Dimroth, T. N. D. Tibbits, M. Niemeyer, F. Predan, P. Beutel, C. Karcher, E. Oliva, G. Siefer, D. Lackner, P. Fuß-Kailuweit, A. W. Bett, R. Krause, C. Drazek, E. Guiot, J. Wasselin, A. Tauzin, and T. Signamarcheix, *Four-Junction Wafer-Bonded Concentrator Solar Cells*, IEEE J. Photovolt. **6**, 343 (2016).
92. F. Dimroth et al., Wafer bonded four-junction GaInP/GaAs//GaInAsP/GaInAs concentrator solar cells with 44.7% efficiency, *Progr. Photovolt.: Res. Appl.* **22**, 277 (2014).
93. R. K. Jones, J. H. Ermer, C. M. Fetzer, and R. R. King, Evolution of multijunction solar cell technology for concentrating photovoltaics, *Jpn. J. Appl. Phys.* **51**, 10ND01 (2012).
94. W. Guter, J. Schöne, S. P. Philipps, M. Steiner, G. Siefer, A. Wekkeli, E. Welser, E. Oliva, A. W. Bett, and F. Dimroth, Current-matched triple-junction solar cell reaching 41.1% conversion efficiency under concentrated sunlight, *Appl. Phys. Lett.* **94**, 223504 (2009).
95. R. M. France, J. F. Geisz, I. García, M. A. Steiner, W. E. McMahon, D. J. Friedman, T. E. Moriarty, C. Osterwald, J. S. Ward, A. Duda, M. Young, and W. J. Olavarria, Quadruple-junction inverted metamorphic concentrator devices, *IEEE J. Photovoltaics* **5**, 432 (2015).
96. A. De Vos and H. Pauwels, On the thermodynamic limit of photovoltaic energy conversion, *Appl. Phys.* **25**, 119 (1981).
97. S. Sze, *Semiconductor Devices: Physics and Technology* (John Wiley & Sons, New York, 1985).
98. H. Morkoç, H. Unlu, and G. Ji, *Principles and Technology of MODFETs* (Wiley, Chichester, 1991), vol. 1 and 2.
99. K. Brennan, K. Hess, J. Y. F. Tang, and G. J. Iafrate, Transient electronic transport in InP under the condition of high-energy electron injection, *IEEE Trans. Electron. Dev.* **30**, 1750 (1983).
100. J. M. Barker, D. K. Ferry, D. D. Koleske, and R. J. Shul, Bulk GaN and AlGaN/GaN heterostructure drift velocity measurements and comparison to theoretical models, *J. Appl. Phys.* **97**, 063705 (2005).

101. S. J. Pearton, F. Ren, A. P. Zhang, and K. P. Lee, Fabrication and performance of GaN electronic devices, *Mater. Sci. Eng.: R: Reports* **30**, 55 (2000).
102. A. Dargys, R. Sedrakyan, and J. Pozela, Measurement of the electron drift velocity in InSb up to fields of 800 V/cm in the presence of impact ionisation, *Phys. Status Solidi A* **45**, 387 (1978).
103. K. Brennan and K. Hess, High field transport in GaAs, InP and InAs, *Solid State Electron.* **27**, 347 (1984).
104. J. S. Blakemore, Semiconducting and other major properties of gallium arsenide, *J. Appl. Phys.* **53**, R123 (1982).
105. J. Becker, E. Fretwurst, and R. Klanner, Measurements of charge carrier mobilities and drift velocity saturation in bulk silicon of <1 1 1> and <1 0 0> crystal orientation at high electric fields, *Solid State Electron.* **56**, 104 (2011).
106. F. Sacconi, A. Di Carlo, P. Lugli, and H. Morkoç, Quasi two-dimensional modeling of GaN-based MODFETs, *Phys. Status Solidi A* **188**, 251 (2001).
107. B. van Zeghbroeck, *Principles of Semiconductor Devices* (http://ece-www.colorado.edu/ bart/book/, University of Colorado, 2011).
108. W. Hafez and M. Feng, Experimental demonstration of pseudomorphic heterojunction bipolar transistors with cutoff frequencies above 600 GHz, *Appl. Phys. Lett.* **86**, 152101 (2005).

Index

For Product Safety Concerns and Information please contact our EU
representative GPSR@taylorandfrancis.com
Taylor & Francis Verlag GmbH, Kaufingerstraße 24, 80331 München, Germany

www.ingramcontent.com/pod-product-compliance
Ingram Content Group UK Ltd.
Pitfield, Milton Keynes, MK11 3LW, UK
UKHW021109180425
457613UK00001B/4

* 9 7 8 9 8 1 4 7 7 4 0 7 9 *